Honda CB250N & CB400N Super Dreams Owners Workshop Manual

by Martyn Meek

With an additional Chapter on the 1981 to 1984 models by Penelope A. Cox

Models covered

CB250 N. 249cc. Introduced February 1978	CB250 NDB. 249cc. Introduced February 1981
CB400 N. 395cc. Introduced May 1978	CB400 NB. 395cc. Introduced February 1981
CB250 NA. 249cc. Introduced November 1979	CB250 NDC. 249cc. Introduced 1982
CB400 NA. 395cc. Introduced November 1979	CB400 NC. 395cc. Introduced 1982
CB250 NB. 249cc. Introduced February 1981	

ISBN 978 0 85696 893 8

(540-4S10)

Haynes Group Limited
Haynes North America, Inc

www.haynes.com

Acknowledgements

Our thanks are due to Honda (UK) Limited for permission to reproduce their line drawings and for supplying the colour transparency of the CB400 NC featured on the front cover. Paul Branson, of Paul Branson Motorcycles Ltd, Yeovil, supplied the Honda CB250 N, and various technical information, used in the preparation of the manual.

Brian Horsfall gave considerable assistance with the strip-down and rebuilding and devised the ingenious methods for overcoming the lack of service tools. Tony Steadman arranged and took the photographs that accompany the text. Mansur Darlington edited the text.

Finally, we would also like to thank the Avon Rubber Company, who kindly supplied advice about tyre fitting; NGK Spark Plugs (UK) Ltd who furnished advice about sparking plug conditions, and Renold Limited who supplied details of replacement chains.

About this manual

The purpose of this manual is to present the owner with a concise and graphic guide which will enable him to tackle any operation from basic routine maintenance to a major overhaul. It has been assumed that any work will be undertaken without the luxury of a well-equipped workshop and a range of manufacturer's service tools.

To this end, the machine featured in the manual was stripped and rebuilt in our own workshop, by a team comprising a mechanic, a photographer and the author. The resulting photographic sequence depicts events as they took place, the hands shown being those of the author and the mechanic.

The use of specialised, and expensive, service tools was avoided unless their use was considered to be essential due to risk of breakage or injury. There is usually some way of improvising a method of removing a stubborn component, provided that a suitable degree of care is exercised.

The author learnt his motorcycle mechanics over a number of years, faced with the same difficulties and using similar facilities to those encountered by most owners. It is hoped that this practical experience can be passed on through the pages of this manual.

Where possible, a well-used example of the machine is chosen for a workshop project, as this highlights any areas which might be particularly prone to giving rise to problems. In this way, any such difficulties are encountered and resolved before the text is written, and the techniques used to deal with them can be incorporated in the relevant Section. Armed with a working knowledge of the machine, the author undertakes a considerable amount of research in order that the maximum amounts of data can be included in the manual.

Each Chapter is divided into numbered sections. Within these Sections are numbered paragraphs. Cross reference throughout the manual is quite straightforward and logical. When reference is made 'See Section 6.10' it means Section 6, paragraph 10 in the same Chapter. If another Chapter were intended, the reference would read, for example, 'See Chapter 2, Section 6.10'. All the photographs are captioned with a section/paragraph number to which they refer and are relevant to the Chapter text adjacent.

Figures (usually line illustrations) appear in a logical but numerical order, within a given Chapter. Fig. 1.1 therefore refers to the first figure in Chapter 1.

Left-hand and right-hand descriptions of the machines and their components refer to the left and right of a given machine when the rider is seated normally.

Motorcycle manufacturers continually make changes to specifications and recommendations, and these, when notified, are incorporated into our manuals at the earliest opportunity. **We take great pride in the accuracy of information given in this manual, but motorcycle manufacturers make alterations and design changes during the production run of a particular motorcycle of which they do not inform us. No liability can be accepted by the authors or publishers for loss, damage or injury caused by any errors in, or omissions from, the information given.**

Contents

Right-hand view of Honda CB250N

Left-hand view of Honda CB250N

Introduction to the Honda CB250N and CB400N Super Dream models

The present Honda empire, which started in a wooden shack in 1947, now occupies a vast modern factory.

The first motorcycle to be imported into the UK in the early 60's, the 250 cc twin 'Dream', was the thin edge of a wedge which has been the Japanese domination of the motorcycle industry. Strange it looked too, to Western eyes, with pressed steel frame, and 'square' styling.

In 1959, Honda commenced road racing in Europe, at the IOM TT races. They came 'to learn, next year to race, maybe', but walked off with the manufacturer's team award. A few years after this derided start, they were to dominate all classes, with such riders as Mike Hailwood, Jim Redman, and the late Tom Phillis and Bob McIntyre, on four, five and six cylinder machines. Even the previously unbeaten Italian multis no longer had things their own way, and were hard put to continue racing under really competitive terms.

Honda withdrew from racing in 1967, when at the top of the tree, not to return again until the 1977 season started. Honda's success in racing has been mirrored in their sales of road going machinery, a range which included models from 49 cc to 1000 cc, and encompasses engine configurations of widely differing types such as single cylinder, transverse six cylinder and even V-twin types.

The Honda CB250N and the CB400N (known also as Super Dreams) were introduced in the middle of 1978 to take a larger slice of the markets that were previously contested by their forerunners, the CB250T amd CB400T (Dreams). Although still on the market in the USA (where they are known as Hawks) the existing Dreams were only available in the UK for a six month period.

Both models are of a similar design, having many interchangeable components. The Super Dreams have what Honda has termed European styling, with a much more sporting profile than their immediate predecessors, including an integral-looking tank and side panel unit, a slimmer seat and new 'tail' hump, lower handlebars, and footrests respositioned rearwards. Mechanically there have also been several changes. Both models now have a six-speed gearbox, modified inlet and exhaust porting and valve timing, redesigned crankcases, and longer silencers. The 400 version now sports a twin disc front brake and a quartz halogen headlamp unit. Reference is made in the text, in the dismantling and reassembly procedures, to the various differences in each model's equipment.

Model dimensions and weight

	CB250N	CB400N
Overall length	2115 mm (83·3 in)	2115 mm (83·3 in)
Overall width	730 mm (28·7 in)	730 mm (28·7 in)
Height	1105 mm (43·5 in)	1105 mm (43·5 in)
Wheelbase	1395 mm (54·9 in)	1390 mm (54·7 in)
Ground clearance	165 mm (6·5 in)	165 mm (6·5 in)
Dry weight	167 kg (367 lb)	171 kg (377 lb)

Safety first!

Professional motor mechanics are trained in safe working procedures. However enthusiastic you may be about getting on with the job in hand, do take the time to ensure that your safety is not put at risk. A moment's lack of attention can result in an accident, as can failure to observe certain elementary precautions.

There will always be new ways of having accidents, and the following points do not pretend to be a comprehensive list of all dangers; they are intended rather to make you aware of the risks and to encourage a safety-conscious approach to all work you carry out on your vehicle.

Essential DOs and DON'Ts

DON'T start the engine without first ascertaining that the transmission is in neutral.

DON'T suddenly remove the filler cap from a hot cooling system – cover it with a cloth and release the pressure gradually first, or you may get scalded by escaping coolant.

DON'T attempt to drain oil until you are sure it has cooled sufficiently to avoid scalding you.

DON'T grasp any part of the engine, exhaust or silencer without first ascertaining that it is sufficiently cool to avoid burning you.

DON'T allow brake fluid or antifreeze to contact the machine's paintwork or plastic components.

DON'T syphon toxic liquids such as fuel, brake fluid or antifreeze by mouth, or allow them to remain on your skin.

DON'T inhale dust – it may be injurious to health (see *Asbestos* heading).

DON'T allow any spilt oil or grease to remain on the floor – wipe it up straight away, before someone slips on it.

DON'T use ill-fitting spanners or other tools which may slip and cause injury.

DON'T attempt to lift a heavy component which may be beyond your capability – get assistance.

DON'T rush to finish a job, or take unverified short cuts.

DON'T allow children or animals in or around an unattended vehicle.

DON'T inflate a tyre to a pressure above the recommended maximum. Apart from overstressing the carcase and wheel rim, in extreme cases the tyre may blow off forcibly.

DO ensure that the machine is supported securely at all times. This is especially important when the machine is blocked up to aid wheel or fork removal.

DO take care when attempting to slacken a stubborn nut or bolt. It is generally better to pull on a spanner, rather than push, so that if slippage occurs you fall away from the machine rather than on to it.

DO wear eye protection when using power tools such as drill, sander, bench grinder etc.

DO use a barrier cream on your hands prior to undertaking dirty jobs – it will protect your skin from infection as well as making the dirt easier to remove afterwards; but make sure your hands aren't left slippery. Note that long-term contact with used engine oil can be a health hazard.

DO keep loose clothing (cuffs, tie etc) and long hair well out of the way of moving mechanical parts.

DO remove rings, wristwatch etc, before working on the vehicle – especially the electrical system.

DO keep your work area tidy – it is only too easy to fall over articles left lying around.

DO exercise caution when compressing springs for removal or installation. Ensure that the tension is applied and released in a controlled manner, using suitable tools which preclude the possibility of the spring escaping violently.

DO ensure that any lifting tackle used has a safe working load rating adequate for the job.

DO get someone to check periodically that all is well, when working alone on the vehicle.

DO carry out work in a logical sequence and check that everything is correctly assembled and tightened afterwards.

DO remember that your vehicle's safety affects that of yourself and others. If in doubt on any point, get specialist advice.

IF, in spite of following these precautions, you are unfortunate enough to injure yourself, seek medical attention as soon as possible.

Asbestos

Certain friction, insulating, sealing, and other products – such as brake linings, clutch linings, gaskets, etc – contain asbestos. *Extreme care must be taken to avoid inhalation of dust from such products since it is hazardous to health.* If in doubt, assume that they *do* contain asbestos.

Fire

Remember at all times that petrol (gasoline) is highly flammable. Never smoke, or have any kind of naked flame around, when working on the vehicle. But the risk does not end there – a spark caused by an electrical short-circuit, by two metal surfaces contacting each other, by careless use of tools, or even by static electricity built up in your body under certain conditions, can ignite petrol vapour, which in a confined space is highly explosive.

Always disconnect the battery earth (ground) terminal before working on any part of the fuel or electrical system, and never risk spilling fuel on to a hot engine or exhaust.

It is recommended that a fire extinguisher of a type suitable for fuel and electrical fires is kept handy in the garage or workplace at all times. Never try to extinguish a fuel or electrical fire with water.

Note: *Any reference to a 'torch' appearing in this manual should always be taken to mean a hand-held battery-operated electric lamp or flashlight. It does **not** mean a welding/gas torch or blowlamp.*

Fumes

Certain fumes are highly toxic and can quickly cause unconsciousness and even death if inhaled to any extent. Petrol (gasoline) vapour comes into this category, as do the vapours from certain solvents such as trichloroethylene. Any draining or pouring of such volatile fluids should be done in a well ventilated area.

When using cleaning fluids and solvents, read the instructions carefully. Never use materials from unmarked containers – they may give off poisonous vapours.

Never run the engine of a motor vehicle in an enclosed space such as a garage. Exhaust fumes contain carbon monoxide which is extremely poisonous; if you need to run the engine, always do so in the open air or at least have the rear of the vehicle outside the workplace.

The battery

Never cause a spark, or allow a naked light, near the vehicle's battery. It will normally be giving off a certain amount of hydrogen gas, which is highly explosive.

Always disconnect the battery earth (ground) terminal before working on the fuel or electrical systems.

If possible, loosen the filler plugs or cover when charging the battery from an external source. Do not charge at an excessive rate or the battery may burst.

Take care when topping up and when carrying the battery. The acid electrolyte, even when diluted, is very corrosive and should not be allowed to contact the eyes or skin.

If you ever need to prepare electrolyte yourself, always add the acid slowly to the water, and never the other way round. Protect against splashes by wearing rubber gloves and goggles.

Mains electricity and electrical equipment

When using an electric power tool, inspection light etc, always ensure that the appliance is correctly connected to its plug and that, where necessary, it is properly earthed (grounded). Do not use such appliances in damp conditions and, again, beware of creating a spark or applying excessive heat in the vicinity of fuel or fuel vapour. Also ensure that the appliances meet the relevant national safety standards.

Ignition HT voltage

A severe electric shock can result from touching certain parts of the ignition system, such as the HT leads, when the engine is running or being cranked, particularly if components are damp or the insulation is defective. Where an electronic ignition system is fitted, the HT voltage is much higher and could prove fatal.

Ordering spare parts

When ordering spare parts for any Honda, it is advisable to deal direct with an official Honda agent, who should be able to supply most items ex-stock. Parts cannot be obtained from Honda (UK) Limited direct; all orders must be routed via an approved agent, even if the parts required are not held in stock.

Always quote the engine and frame numbers in full, and colour when painted parts are required.

The frame number is located on the side of the steering head, and the engine number is stamped on the crankcase below the right-hand carburettor.

Use only parts of genuine Honda manufacture. Pattern parts are available, some of which originate from Japan, but in many instances they may have an adverse effect on performance and/or reliability. Honda do not operate a 'service exchange' scheme.

Some of the more expendable parts such as sparking plugs, bulbs, tyres, oils and greases etc., can be obtained from accessory shops and motor factors, who have convenient opening hours, and can often be found not far from home. It is also possible to obtain parts on a Mail Order basis from a number of specialists who advertise regularly in the motorcycle magazines.

Location of engine number

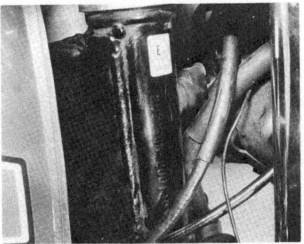

Location of frame number

Routine Maintenance

Periodic routine maintenance is a continuous process which should commence immediately the machine is used. The object is to maintain all adjustments and to diagnose and rectify minor defects before they develop into more extensive, and often more expensive, problems.

It follows that if the machine is maintained properly, it will both run and perform with optimum efficiency, and be less prone to unexpected breakdowns. Regular inspection of the machine will show up any parts which are wearing, and with a little experience, it is possible to obtain the maximum life from any one component, renewing it when it becomes so worn that it is liable to fail.

Regular cleaning can be considered as important as mechanical maintenance. This will ensure that all the cycle parts are inspected regularly and are kept free from accumulations of road dirt and grime.

Cleaning is especially important during the winter months, despite its appearance of being a thankless task which very soon appears pointless. On the contrary, it is during these months that the paintwork, chromium plating, and the alloy casing suffer the ravages of abrasive grit, rain and road salt. A couple of hours spent weekly on cleaning the machine will maintain its appearance and value, and highlight small points, like chipped paint, before they become a serious problem.

The various maintenance tasks are described under their respective mileage and calendar headings, and are accompanied by diagrams and photographs, where pertinent.

It should be noted that the intervals between each maintenance task serve only as a guide. As the machine gets older, or if it is used under particularly arduous conditions, it is advisable to reduce the period between each check.

For ease of reference, most service operations are described in detail under the relevant heading. However, if further general information is required, this can be found under the pertinent Section heading and Chapter in the main text.

No special tools are required for the normal routine maintenance tasks. The tools contained in the kit supplied with every new machine will prove adequate for each task, but if they are not available, the tools found in the average household should suffice.

Additional items, such as a good quality socket set, and an impact driver may be added to the list, as can a small multimeter which is invaluable for diagnosing electrical faults.

Weekly or every 300 miles (500 km)

---

1 Topping up the engine/transmission oil

Unscrew the combined filler plug and dipstick, which is situated in the primary drive cover. Wipe off the dipstick and place it in position, but do not screw it home. Withdraw it and note the reading. If necessary, top up the oil level to bring it to the MAX position. Honda recommend the use of a good quality SAE 10W/40 engine oil. If this is unavailable SAE 15W/40 or 20W/50 oil may be used.

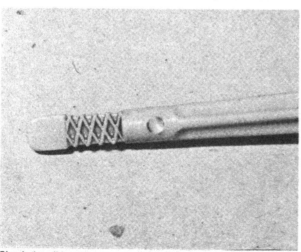

Check the oil level by means of the dipstick and ...

... replenish, if required, with the correct grade of oil

2   Tyre pressures

Check the tyre pressures with a pressure gauge which is known to be accurate. It is preferable to purchase a gauge so that inaccurate readings from filling station gauges are not encountered. Keep the valve free from dust or mud and always replace the dust cap. Check the pressures when the tyres are cold. If the machine has travelled a number of miles, the tyres will have become hot, and the pressure will have increased. A false reading will therefore result.

Tyre pressures
Front   24 psi (1·75 kg/cm²)
*Rear   32 psi (2·25 kg/cm²)

* Increase the pressure in the rear tyre by 4 psi (0·25 kg/cm²) when carrying a pillion passenger or travelling at continuous high speed.

3   Battery electrolyte level

Access to the Yuasa battery is gained after removing the right-hand side panel. The electrolyte level can be checked visually through the battery's transparent case. Make sure that the level in each cell is between the minimum and maximum lines on the battery case and that the vent pipe has not become pinched or obstructed. The transparent case also makes it possible for a quick check on the condition of the battery plates and separators.

Unless acid is spilt, as may occur if the machine falls over, the electrolyte should always be topped up with distilled water to restore the correct level. If acid is spilt on any part of the machine, it should be neutralised with an alkali such as washing soda, and washed away with plenty of water, otherwise serious corrosion will occur. Top up with sulphuric acid, only in the event of spillage.

4   Control cables

Visually check each cable end for fraying or broken strands. If necessary, adjust to take up any excess free play. Apply a few drops of oil to the exposed inner portion of each control cable and allow it to run down into the outer sheathing. This will prevent the cables sticking or drying up.

5   Rear chain lubrication and adjustment

To prolong the life of the chain, regular lubrication and adjustment is essential.

Intermediate lubrication should take place weekly with the chain in situ. Application of one of the aerosol chain lubricants is ideal for this purpose. Ordinary engine oil can be used, though due to the speed at which it is flung off, its effectiveness is limited.

Adjust the chain after lubrication so that there is approximately 15 – 20 mm (⅝ – ¾ in) free movement in the middle of the lower run. Always check with the chain at its tightest point, (it will rarely wear evenly during service), with the machine off its stand and with one person sitting on the rear of the seat.

Adjustment is accomplished after placing the machine on the centre stand and slackening the wheel nut, so that the wheel can be drawn backwards by means of the drawbolt adjusters in the fork ends. The torque arm nuts and the caliper bracket nut must also be slackened during this operation. Adjust the drawbolts an equal amount to preserve wheel alignment. The fork ends are clearly marked with a series of parallel lines above the adjusters, to provide a simple visual check.

6   Safety check

In addition to the check on control cables mentioned earlier, check for loose nuts and fittings, and examine the tyres for wear or damage. Pay particular attention to the sidewalls, looking for any possible splits or tears. Remove any stones or other objects which may have become trapped between the treads. This is particularly important in the case of the front tyre, where rapid deflation will cause total loss of control.

7   Legal check

Ensure that the horn, lights, speedometer and indicators all function correctly. Do not forget to check the efficiency of the brakes.

Six weekly or every 1800 miles (3000 km)

Carry out the checks listed under the weekly/300 mile heading and then complete the following:

1   Engine/transmission oil

With the engine warm, to facilitate thorough draining, the machine should be placed on its centre stand and the crankcase sump contents drained into a suitable container. The oil capacity is approximately 3.0 litres (5.3 pints), so ensure that a container of adequate capacity is to hand before removing the drain plug and filler cap/dipstick. The drain plug is located in the lower right-hand wall of the crankcase, below the primary drive cover. When the oil has drained completely, refit the drain plug. Check that the sealing ring is in good condition, renewing it if necessary. Refill the engine with 2.2 litres (3.9 pints) of SAE 10W/30 engine oil. SAE 15W/40 or 20W/50 may also be used. Allow the level to settle and then top up, as indicated by the reading on the dipstick.

Check tyre pressures

Battery electrolyte level can be seen through translucent case

Drain the engine oil

Three monthly or every 3600 miles (6000 km)

Complete the tasks listed under the previous headings and then complete the following:

1 Renewing the oil filter

The oil filter should be renewed at every second oil change. The filter is of the corrugated paper type and **cannot** be cleaned; it must be replaced by a new filter.

After draining the engine oil reposition the drain pan under the finned filter chamber cover in the base of the crankcase. Unscrew the central bolt and lower the bolt, chamber cover and filter element from position. The filter element is a push fit on the hollow bolt. Note the washer and spring fitted below the filter. These must not be omitted on reassembly.

Before replacing the oil filter element check that the sealing O-ring in the chamber cover groove is in good condition and positioned correctly. Refit the filter, chamber cover and bolt by reversing the dismantling procedure. Note the forked projection on the chamber cover. This must locate with the lug projecting from the crankcase, if the chamber is positioned correctly.

2 Air filter element cleaning

The air filter element must be removed at regular intervals for cleaning to prevent excessive clogging and a resultant increase in fuel consumption and loss of engine performance. If the machine is used in particularly dusty conditions, the maintenance periods should be shortened accordingly. To gain access to the filter element, the dualseat must first be removed. The seat is secured by two spring loaded catches at the rear of the seat below the seat pan.

The filter box is located below the nose of the dualseat and is closed by a cover held by three screws. Remove the screws and lift the cover away. Lift out the element frame, which is located at the rear by two lugs projecting from the casing, and then remove the element. The element is made of synthetic foam sheeting, impregnated with oil, and should be cleaned in a high flashpoint solvent. Petrol may be used as an alternative, but the element **must** be allowed to dry thoroughly before the application of new oil. A petrol-wet element will constitute a fire hazard if blow-back occurs.

When the element is dry, it should be impregnated with clean gear oil (SAE 80-90) and squeezed to remove the excess. **Do not** wring out the element because this may cause damage to the fabric, necessitating renewal. If the air filter element is perforated or perished it should be renewed forthwith.

On no account run without the air cleaners attached, or with the element missing. The jetting of the carburettors takes

into account the presence of the air cleaner and engine performance will be seriously affected if this balance is upset.

To replace the element, reverse the dismantling procedure. Give a visual check to ensure that the inlet hoses are correctly located and not kinked, split or otherwise damaged. Check that the air cleaner cases are free from splits or cracks.

3 Sparking plugs

Remove both sparking plugs for cleaning and re-gapping. Before cleaning, compare the plugs with those shown in the sparking plug condition chart in the Ignition Chapter, to obtain an indication of the running condition of each cylinder. Clean both plugs, using a wire brush – the type used for cleaning suede shoes is ideal – and carefully clean the electrode faces with a fine swiss file or magneto file.

Check the electrode gap with a feeler gauge and adjust, if necessary, to give a clearance of 0·7 mm (0·028 in). Always set the gap by bending the outer (earth) electrode, ensuring that the two electrode faces are kept square. On no account should the centre electrode be bent as this will only succeed in cracking the ceramic insulator, rendering the plug useless. The sparking plugs must be of the correct length (reach) and grade.

When refitting the sparking plugs, check that the HT leads and suppressor caps are in good condition.

4 Cam chain tension

The cam chain can be adjusted either dynamically or statically.

To perform the dynamic adjustment, start the engine and allow it to run at tickover speed. Slacken the cam chain tensioner locknut (located at the rear of the cylinder block) and then retighten it. The tensioner will tension the chain automatically, when the locknut is slackened. Although the dynamic method is quicker to perform it is recommended that if there is any doubt about the tensioner operation, the static procedure is followed.

With the engine stopped, remove the engine left-hand cover and rotate the crankshaft anti-clockwise until the rotor T mark and fixed index mark align. At this point slacken the tensioner locknut and retighten it. In some cases the tensioner may have become stuck and normal adjustment will have no effect on chain tension. If this is suspected remove the cylinder head cover (engine stopped) and gently press down on the top of the tensioner blade, with the locknut slackened; this should serve to free the tensioner.

5 Valve clearance (rocker arm) adjustment

To gain access to the rocker arms so that adjustment can take place, the petrol tank, seat and rocker cover must be removed. In addition, removal of the sparking plugs will aid rotation of the engine during the clearance checking operation.

When removing the petrol tank, remember to turn the fuel tap to the 'OFF' position before removing the feed pipe.

The rocker cover is held by two special sleeve nuts under which fit conical sealing plugs. Remove the nuts and lift the cover away. Place the machine in top gear, and by turning the rear wheel rotate the engine until the right-hand cylinder is at TDC. This position can be found with ease by placing the milled slot in the end of the camshaft in the 12 o'clock position. The correct clearances, with the engine cold, are as follows:

	CB250 N	CB400 N
Inlet	0.12 mm (0.005 in)	0.10 mm (0.004 in)
Exhaust	0.16 mm (0.006 in)	0.14 mm (0.005 in)

If adjustment is required, slacken the locknut securing the adjuster screw. Rotate the screw until the feeler gauge is a light sliding fit, and then without allowing the adjuster screw to turn, tighten the locknut fully. Re-check the gap after tightening the locknut. When the clearances are correct, rotate the engine through 180° and check, and if necessary adjust, the valve clearances on the left-hand cylinder. Refit the rocker cover, checking that the sealing ring and sealing plugs are in good con-

Lift out the air filter element

Feeler gauge should be a light sliding fit

dition, and fit and tighten the nuts. Replace the petrol tank and seat.

6 Carburettor synchronisation

For the best possible performance it is imperative that the carburettors are working in perfect harmony with each other. At any given throttle opening if the carburettors are not synchronised, not only will one cylinder be doing less work but it will also in effect have to be 'carried' by the other cylinder. This will reduce the performance considerably.

It is essential to use a vacuum gauge set consisting of two separate dial gauges, one of each being connected to each carburettor by means of a special adaptor tube.

Because of the prohibitive cost of a set of the required gauges it is suggested that the machine be returned to a Honda Service Agent for synchronisation and adjustment to be carried out. If the vacuum gauge set is available, refer to Chapter 2, Sections 7 and 8, for the relevant instructions.

Six monthly or every 7200 miles (12 000 km)

Carry out the checks listed under the preceding routine maintenance headings and then complete the following:

Cam chain adjusting nut

1 Fuel filter cleaning

A filter is fitted to the petrol tap, which projects into the petrol tank, and helps prevent foreign matter from entering the fuel system. To gain access to the filter, the petrol must first be drained. Release the tension on the petrol pipe by pinching together the ears of the securing spring clip and detach the pipe. Substitute the feed pipe for a suitable length of tubing, to allow the petrol to be drained into a clean container.

The tap is secured by a gland nut. Slacken the nut fully and withdraw the tap, complete with the filter. The filter may be slipped off the stand pipe for cleaning. This should be done in clean petrol, using a soft brush to dislodge any trapped sediment or rust flakes. If the screen is perforated, it should be renewed.

Re-install the tap by reversing the dismantling procedure. As a precaution against leakage, apply a small amount of sealing compound to the tap boss threads before the tap is refitted.

Nine monthly or every 10 000 miles (18 000 km)

Carry out the checks listed under the foregoing headings and complete the following:

1 Balance weight drive chain adjustment

At this routine maintenance interval the drive chain which drives the balance weights from the crankshaft should be adjusted. This operation should also be carried out if the engine begins to get mechanically noisy in the crankcase area. The balance chain is adjusted by means of an eccentric shaft upon which the front balance weight is mounted. To gain access to the shaft, removal of a slotted cap at the front of the right-hand side crankcase cover is required. The balance weight shaft is located forward of the primary drive pinion and is fitted with a quadrant which limits the movement of the shaft. Slacken the nut on the stud which passes through the elongated slot in the quadrant. When this self-locking nut is loosened the balancer shaft will automatically move, due to its spring loading, to provide correct chain tension. The nut may then be retightened.

If, when the nut is loosened, the quadrant moves hard up against the stud, the quadrant must be repositioned on the end of the splined shaft. This will require removal of the primary drive cover, as follows. Drain the engine oil by removing the drain plug in the crankcase and remove the kickstart lever from the splined shaft.

The lever is held by a single pinch bolt. Detach the clutch cable at the actuating lever after slackening off the adjuster screw. Remove the right-hand forward footrest.

Slacken the primary drive cover retaining screws evenly, in a diagonal sequence, to prevent warpage. Remove the screws and, disconnect the tachometer drive cable from the front of the cover. Lift the cover away.

Remove the locknut and shaft end nut. Pull off the quadrant and reposition it on the splines so that the stud is lying centrally in the quadrant slot. Note the punch mark on the shaft end. If the mark is lower than the 3 o'clock position the chain is badly worn and requires renewal (see the accompanying illustration). Fit and tighten the shaft nut, followed by the locknut. Replace the primary drive cover, using a new gasket, and refit the related components by reversing the dismantling procedure.

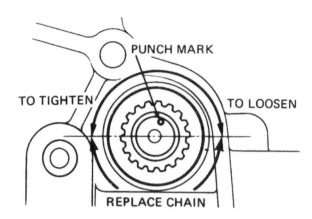

Balance chain requires renewal if punch mark is below the horizontal line

---

Yearly or every 14 500 miles (24 000 km)

Complete all the checks listed under the preceding headings and then carry out the following:

1   Front fork fluid
See Chapter 4, Sections 2 and 3, paragraphs 1–3.

2   Brake lining wear
Remove the rear wheel and inspect the brake shoe linings for wear. See Chapter 5, Section 12.

3   Wheel bearings
Inspect and regrease the wheel bearings. See Chapter 5, Sections 10 and 15.

4   Hydraulic fluid
The hydraulic fluid used in disc brake systems is a specially formulated liquid capable of transmitting hydraulic pressure efficiently. Over a period of time it ages, or degrades, to the stage where it cannot be relied upon to do its job effectively. This is mainly due to its hygroscopic properties, that is, it attracts and absorbs moisture from the air around it. The DOT or SAE rating of a fluid means that it has been manufactured to operate under the normal hot conditions caused by braking, and that it will not boil in use. As it becomes degraded, however, the boiling point falls until minute air bubbles form in the fluid. Ultimately, this results in a drop in braking efficiency. To avoid this possibility, it is advised that the brake fluid be changed annually, and also every time the brakes are bled. This is accomplished by following the brake bleeding instructions described in Chapter 5, ensuring that all the old fluid is drained off and replaced by new fluid. This process will also remove any traces of abrasive contaminant which might otherwise cause wear in the seals and pistons. It is a good idea at this stage, to dismantle and examine the various hydraulic components as a safeguard against possible failure. This aspect is also covered in detail in Chapter 5.

Routine maintenance – additional information

It may well have been noticed that specific mention has not been made of various items such as brake pad inspection and clutch and brake adjustment etc. These cannot be tied to a specific mileage or interval of time, and should be performed as the occasion demands. Likewise, no mention has been made of checking the master cylinder reservoir level. In normal service, the level will remain virtually unchanged, in which case unnecessary inspection may introduce contaminants. Conversely, in the event of a failure of the system, the contents of the reservoir will be lost within a few applications of the brake. As in the case of the various cable adjustments, the interval must remain at the discretion of the owner.

Additionally, it should be noted here that any part of the machine not specifically dealt with in this Section should not be thought of as immune to wear or failure.

The experienced owner will be aware of a continuous feedback of information from the machine whilst in use, any small points worthy of investigation making themselves known. In time, a cursory glance over the vital components will become second nature.

Use this Maintenance Section as a guide, but avoid the temptation to think that any one component, once attended to, may be completely ignored for the next 6000 miles. Maintenance is as much a part of motorcycling as is the riding of the motorcycle itself.

Quick glance
maintenance adjustments and capacities

Engine/transmission oil capacity:
At oil change . . . . 2.2 litre (3.9 pint)
At oil and filter change . . . . 2.3 litre (4.0 pint)

Front fork legs (per leg) . . . . 140 cc (4.9 fl oz)

Sparking plug gap . . . . 0·6 – 0·7 mm (0·024 – 0·028 in)

Tyre pressures:
Front . . . . 24 psi (1·75 kg/cm^2)
*Rear . . . . 32 psi (2·25 kg/cm^2)

Increase pressure in rear tyre by 4 psi (0·25 kg/cm^2) when carrying a passenger or travelling at continuous high speed.

Recommended lubricants

Engine/gearbox . . . . SAE 10W/40, 15W/40 or 20W/50

Front forks . . . . Automatic transmission fluid (ATF)

Final drive chain . . . . Linklyfe and aerosol chain lubricant

Greasing points . . . . High melting point grease

Wheel bearings . . . . High melting point grease

Working conditions and tools

When a major overhaul is contemplated, it is important that a clean, well-lit working space is available, equipped with a workbench and vice, and with space for laying out or storing the dismantled assemblies in an orderly manner where they are unlikely to be disturbed. The use of a good workshop will give the satisfaction of work done in comfort and without haste, where there is little chance of the machine being dismantled and reassembled in anything other than clean surroundings. Unfortunately, these ideal working conditions are not always practicable and under these latter circumstances when improvisation is called for, extra care and time will be needed.

The other essential requirement is a comprehensive set of good quality tools. Quality is of prime importance since cheap tools will prove expensive in the long run if they slip or break when in use, causing personal injury or expensive damage to the component being worked on. A good quality tool will last a long time, and more than justify the cost.

For practically all tools, a tool factor is the best source since he will have a very comprehensive range compared with the average garage or accessory shop. Having said that, accessory shops often offer excellent quality tools at discount prices, so it pays to shop around. There are plenty of tools around at reasonable prices, but always aim to purchase items which meet the relevant national safety standards. If in doubt, seek the advice of the shop proprietor or manager before making a purchase.

The basis of any tool kit is a set of open-ended spanners, which can be used on almost any part of the machine to which there is reasonable access. A set of ring spanners makes a useful addition, since they can be used on nuts that are very tight or where access is restricted. Where the cost has to be kept within reasonable bounds, a compromise can be effected with a set of combination spanners – open-ended at one end and having a ring of the same size on the other end. Socket spanners may also be considered a good investment, a basic $3/8$ in or $1/2$ in drive kit comprising a ratchet handle and a small number of socket heads, if money is limited. Additional sockets can be purchased, as and when they are required. Provided they are slim in profile, sockets will reach nuts or bolts that are deeply recessed. When purchasing spanners of any kind, make sure the correct size standard is purchased. Almost all machines manufactured outside the UK and the USA have metric nuts and bolts, whilst those produced in Britain have BSF or BSW sizes. The standard used in USA is AF, which is also found on some of the later British machines. Others tools that should be included in the kit are a range of crosshead screwdrivers, a pair of pliers and a hammer.

When considering the purchase of tools, it should be remembered that by carrying out the work oneself, a large proportion of the normal repair cost, made up by labour charges, will be saved. The economy made on even a minor overhaul will go a long way towards the improvement of a toolkit.

In addition to the basic tool kit, certain additional tools can prove invaluable when they are close to hand, to help speed up a multitude of repetitive jobs. For example, an impact screwdriver will ease the removal of screws that have been tightened by a similar tool, during assembly, without a risk of damaging the screw heads. And, of course, it can be used again to retighten the screws, to ensure an oil or airtight seal results. Circlip pliers have their uses too, since gear pinions, shafts and similar components are frequently retained by circlips that are not too easily displaced by a screwdriver. There are two types of circlip pliers, one for internal and one for external circlips. They may also have straight or right-angled jaws.

One of the most useful of all tools is the torque wrench, a form of spanner that can be adjusted to slip when a measured amount of force is applied to any bolt or nut. Torque wrench settings are given in almost every modern workshop or service manual, where the extent to which a complex component, such as a cylinder head, can be tightened without fear of distortion or leakage. The tightening of bearing caps is yet another example. Overtightening will stretch or even break bolts, necessitating extra work to extract the broken portions.

As may be expected, the more sophisticated the machine, the greater is the number of tools likely to be required if it is to be kept in first class condition by the home mechanic. Unfortunately there are certain jobs which cannot be accomplished successfully without the correct equipment and although there is invariably a specialist who will undertake the work for a fee, the home mechanic will have to dig more deeply in his pocket for the purchase of similar equipment if he does not wish to employ the services of others. Here a word of caution is necessary, since some of these jobs are best left to the expert. Although an electrical multimeter of the AVO type will prove helpful in tracing electrical faults, in inexperienced hands it may irrevocably damage some of the electrical components if a test current is passed through them in the wrong direction. This can apply to the synchronisation of twin or multiple carburettors too, where a certain amount of expertise is needed when setting them up with vacuum gauges. These are, however, exceptions. Some instruments, such as a strobe lamp, are virtually essential when checking the timing of a machine powered by CDI ignition system. In short, do not purchase any of these special items unless you have the experience to use them correctly.

Although this manual shows how components can be removed and replaced without the use of special service tools (unless absolutely essential), it is worthwhile giving consideration to the purchase of the more commonly used tools if the machine is regarded as a long term purchase Whilst the alternative methods suggested will remove and replace parts without risk of damage, the use of the special tools recommended and sold by the manufacturer will invariably save time.

Chapter 1 Engine, clutch and gearbox

Contents

Specifications

Engine

	CB250N
Type	Twin-cylinder, air-cooled, 4-stroke
Bore	62.0 mm (2.441 in)
Stroke	41.4 mm (1.630 in)
Capacity	249 cc (15.3 cu in)
Compression ratio	9.4 : 1

Pistons and rings

Piston OD (standard)	61.97-61.99 mm (2.440-2.441 in)
Ring end gap	0.10-0.30 mm (0.004-0.012 in)
Service limit	0.50 mm (0.020 in)
Ring side clearance:	
Top ring	0.03-0.06 mm (0.001-0.002 in)
Service limit	0.10 mm (0.004 in)
Second ring	0.02-0.05 mm (0.0008-0.0020 in)
Service limit	0.10 mm (0.004 in)
Oil control ring	0.015-0.045 mm (0.0006-0.0018 in)
Service limit	0.10 mm (0.004 in)
Piston oversizes available	+ 0.25 mm (0.010 in) + 0.50 mm (0.020 in)
	+ 0.75 mm (0.030 in) and + 1.0 mm (0.040 in)

Cylinder block

Cylinder bore standard	62.00-62.01 mm (2.4409-2.4413)
Service limit	62.10 mm (2.444 in)
Maximum ovality	0.10 mm (0.004 in)
Piston/cylinder clearance	0.10 mm (0.004 in)

Valves and springs

Valve seat angle	45°
Valve stem diameter (min):	
Inlet	5.44 mm (0.214 in)
Exhaust	6.54 mm (0.257 in)
Valve stem/guide clearance:	
Inlet	0.10 mm (0.004 in)
Exhaust	0.10 mm (0.004 in)
Valve spring free length (min):	
Inlet outer	44.5 mm (1.75 in)
Inlet inner	36.0 mm (1.42 in)
Exhaust outer	43.9 mm (1.73 in)
Exhaust inner	41.0 mm (1.61 in)
Valve clearance (cold):	
Inlet	0.12 mm (0.005 in)
Exhaust	0.16 mm (0.006 in)

Camshaft and rocker arms

Camshaft lobe minimum height	
Inlet	36.9 mm (1.45 in)
Exhaust	36.9 mm (1.45 in)
Camshaft journal/bearing clearance (max):	
Centre journal	0.23 mm (0.0090 in)
Outer journals	0.20 mm (0.0079 in)

Valve timing

Inlet opens	10° BTDC
Inlet closes	30° ABDC
Exhaust opens	40° BBDC
Exhaust closes	5° ATDC

Crankshaft

Main bearing/journal clearance	0.020-0.045 mm (0.0008-0.0018 in)
Service limit	0.08 mm (0.003 in)
Big-end bearing/journal clearance	0.020-0.044 mm (0.0008-0.0017 in)
Service limit	0.08 mm (0.003 in)
Big-end side float	0.05-0.25 mm (0.002-0.010 in)
Service limit	0.35 mm (0.014 in)

Clutch

Type	Wet, multi-plate
No of plates:	
Friction type A	6
Friction type B	1
Plain	6
Friction plate thickness:	
Type A	2.7 mm (0.106 in)
Service limit	2.3 mm (0.090 in)
Type B	3.0 mm (0.118 in)
Service limit	2.6 mm (0.102 in)
Spring free length	39.0 mm (1.535 in)
Service limit	37.5 mm (1.476 in)

Gearbox

Type	6-speed, constant-mesh
Gear ratios:	
1st gear	2.733 : 1
2nd gear	1.947 : 1
3rd gear	1.545 : 1
4th gear	1.280 : 1
5th gear	1.074 : 1
6th gear	0.930 : 1
Primary drive ratio	3.125 : 1

Engine

CB400N

Type	Twin-cylinder, air-cooled, 4-stroke
Bore	70.5 mm (2.776 in)
Stroke	50.6 mm (1.992 in)
Capacity	395 cc (24.1 cu in)
Compression ratio	9.3 : 1

Pistons and rings

Piston OD (standard)	70.47-70.49 mm (2.774-2.775 in)
Ring end gap:	
Top and second ring	0.2-0.4 mm (0.008-0.016 in)
Service limit	0.6 mm (0.024 in)
Oil control ring	0.2-0.9 mm (0.008-0.035 in)
Service limit	1.10 mm (0.043 in)
Ring side clearance:	
Top ring	0.03-0.06 mm (0.001-0.002 in)
Service limit	0.10 mm (0.004 in)
Second ring	0.025-0.055 mm (0.0009-0.0022 in)
Service limit	0.10 mm (0.004 in)
Piston oversizes available	+ 0.25 mm (0.010 in), + 0.050 mm (0.020 in) + 0.75 mm (0.030 in) and + 1.0 mm (0.040 in)

Cylinder block

Cylinder bore standard	70.50-70.51 mm (2.775-2.776 in)
Service limit	70.60 mm (2.78 in)
Maximum ovality	0.10 mm (0.004 in)
Piston/cylinder clearance	0.10 mm (0.004 in)

Valves and springs

Valve seat angle	45°
Valve stem diameter (min):	
Inlet	5.44 mm (0.214 in)
Exhaust	6.54 mm (0.257 in)
Valve stem/guide clearance:	
Inlet	0.10 mm (0.004 in)
Exhaust	0.10 mm (0.004 in)
Valve spring free length (min):	
Inlet outer	44.5 mm (1.75 in)
Inlet inner	36.0 mm (1.42 in)
Exhaust outer	43.9 mm (1.73 in)
Exhaust inner	41.0 mm (1.61 in)
Valve clearance (cold):	
Inlet	0.10 mm (0.004 in)
Exhaust	0.14 mm (0.005 in)

Camshaft and rocker arms

Camshaft lobe minimum height	
Inlet	37.180 mm (1.4638 in)
Exhaust	37.213 mm (1.4651 in)
Camshaft journal/bearing clearance (max):	
Centre journal	0.23 mm (0.009 in)
Outer journals	0.20 mm (0.008 in)

Valve timing

Inlet opens	10° BTDC
Inlet closes	35° ABDC
Exhaust opens	40° BBDC
Exhaust closes	10° ATDC

Crankshaft

Main bearing/journal clearance	0.020-0.045 mm (0.0008-0.0018 in)
Service limit	0.08 mm (0.003 in)
Big-end bearing/journal clearance	0.020-0.044 mm (0.0008-0.0017 in)
Service limit	0.08 mm (0.003 in)
Big-end side float	0.05-0.25 mm (0.002 to 0.010 in)
Service limit	0.35 mm (0.014 in)

Clutch

Type	Wet, multi-plate
No. of plates: Friction type A	6
Friction type B	1
Plain	6
Friction plate thickness:	
Type A	2.7 mm (0.106 in)
Service limit	2.3 mm (0.090 in)
Type B	3.0 mm (0.118 in)
Service limit	2.6 mm (0.102 in)
Spring free length	42.75 mm (1.863 in)
Service limit	41.25 mm (1.624 in)

Gearbox

Type	6-speed, constant-mesh
Gear ratios:	
1st gear	2.733 : 1
2nd gear	1.947 : 1
3rd gear	1.545 : 1
4th gear	1.280 : 1
5th gear	1.074 : 1
6th gear	0.931 : 1
Primary drive ratio	3.125 : 1

1 General description

The engine fitted to the Honda 250N and 400N models is a single overhead camshaft vertical parallel twin. The 360° webbed crankshaft, which runs on four shell main bearings is balanced by two rotating weights, one of which lies forwards of the crankshaft and one aft. The weights are timed and driven by a chain, driven from a sprocket lying between the two centre bearings and next to the camshaft drive sprocket. The camshaft drive chain, which is of the inverted Hy-Vo type, passes through a tunnel between the cylinders and so to the camshaft which is mounted on the cylinder head. To make full use of the relatively small combustion chamber, each cylinder has three valves; two inlet and one exhaust. The use of two inlet valves, instead of a large single valve giving similar breathing capabilities, increases cylinder filling during the induction stroke, but without the penalty of increased component weight. This allows a higher safe maximum engine speed without the risk of valve bounce, enabling full use to be made of the advantages already gained by improved induction. All engine castings are manufactured from aluminium alloy, the cylinder block being fitted with dry steel liners. The crankcases, which house the crankshaft and gearbox components, separate horizontally to facilitate easy dismantling and reassembly.

The crankshaft is secured by a complex aluminium casting bolted to the underside of the crankcase upper half. This casting carries the lower half of each main bearing shell and the rear balance shaft, and also serves as a mounting for the oil pump and strainer unit.

Wet sump lubrication is supplied by a trochoid oil pump driven by a chain from the crankshaft. The oil is picked up from the reservoir in the sump through a mesh strainer, and is then passed through a paper filter element to all working parts of the engine.

Both models utilize a traditional constant-mesh, 6-speed gearbox driven from the engine via spur primary drive gears and a multi-plate clutch.

On CB250 N and 400 N models starting is effected either by kickstart or electric starter motor. The CB250 NA and 400 NA models, however, are not equipped with a kickstarter and rely entirely on the electric means of starting. Electrical power is generated by a crankshaft-mounted 12-volt alternator which incorporates the source coil and timing device for the CDI (capacitor discharge ignition) system. The CDI system dispenses entirely with the contact breakers, eliminating the need for maintenance and preserving the long-term accuracy of the ignition timing.

2 Operations with engine/gearbox in frame

It is not necessary to remove the engine unit from the frame in order to dismantle the following items:
1 Right and left-hand crankcase covers and final drive sprocket cover.
2 Clutch assembly and gear selector components (external).
3 Oil pump and filter.
4 Alternator and starter motor.
5 Cylinder head and cylinder head cover.
6 Cylinder block, pistons and rings.

3 Operations with engine/gearbox unit removed from frame

As previously described, the crankshaft and gearbox assemblies are housed within a common casing. Any work carried out on either of these two major assemblies will necessitate removal of the engine from the frame so that the crankcase halves can be separated.

4 Removing the engine/gearbox unit

1 Place the machine on its centre stand, making sure that it is

standing firmly.

2 Although by no means essential it is useful to raise the machine a number of feet above floor level by placing it on a long bench or horizontal ramp. This will enable most of the work to be carried out in an upright position.

3 Place a suitable receptacle below the crankcase and drain off the engine oil. The sump plug lies just forward and to the right of the oil filter housing. The oil will drain faster if the engine has been warmed up previously, thereby heating and thinning the oil. Approximately 3 litres (5.5 pints) should drain out. Undo the oil filter chamber bolt and remove the chamber and filter.

4 The dualseat is secured by two spring loaded catches, one of which is placed each side of the seat, towards the rear. Before depressing the catches and lifting the seat away, free the securing link from the lock fitted to the left-hand side of the frame. Detach both side covers from the frame. Each is a push fit, secured by projections locating in grommets in the frame.

5 To prevent the risk of short circuits during subsequent dismantling, the battery should be disconnected. This may be done by prising back the protecting boot from the accessible terminal (positive) and detaching the main cable. If the machine is to be out of service for an extended length of time, the battery should be removed from the machine and given a trickle charge at approximately one month intervals. To free the battery, detach the securing strap and pull the battery out of the carrier, to gain access to the Earth (negative) terminal. Disconnect the earth lead and the breather tube, and lift the battery away.

6 Turn the fuel tap to the 'OFF' position and remove the fuel lines from either the tap unions or the carburettor float chamber unions. The petrol tank is secured at the rear by a single bolt and is supported at the front by rubber buffers attached to the frame top tube, which locate with cups on the underside of the tank. After removing the bolt, slide the tank rearwards, until the cups clear the rubbers, and then lift the tank off the frame.

7 Slacken off fully the exhaust pipe/balance box and balance box/silencer clamps. Then pull the clamps off the box stubs, to clear the joints. Remove the two nuts holding each exhaust port flange and slide the flanges off the studs. The two exhaust pipes may be detached as individual components. The silencers are retained on brackets held to the forged aluminium footrest plates by a single bolt each. The bolt also secures the pillion footrest. After removal of the silencers, the balance box, which is suspended from the crankcase by two bolts, may be detached. The box is surprisingly heavy, so be prepared to support the weight when removing the bolts. Removal of the chrome guard from the right-hand side of the box may be required, to allow access to the support bolt.

8 Pull back the rubber boot from the lower end of the clutch cable so that the rearmost locknut may be loosened and run off the thread. Slacken the nut on the adjuster fully, to give plenty of slack so that the inner cable may be detached from the operating arm. The slot in the arm end may require opening slightly with a screwdriver to allow the cable to be detached. Pull the complete cable out of the anchor bracket. Detach the tachometer drive cable from the crankcase right-hand cover, after unscrewing the locating bolt. Refit the bolt to avoid loss.

9 Disconnect the alternator leads at the two block connectors and two snap connectors to the rectifier/regulator unit. Similarly, disconnect the oil pressure switch lead and neutral indicator lead. Prise off the rubber boot from the starter motor terminal and disconnect the starter cable.

10 Disconnect the breather tube from the unions on the cylinder head and air filter box. The tube is secured at each end by a spring clip. Removal of the carburettors is not strictly necessary when preparing for engine removal. It is suggested, however, that the carburettors are removed to increase clearance both for the removal and re-installation of the engine. Because of the limited space between the rear of the cylinder head and the front of the air filter box, carburettor removal requires some care and should be carried out in the following sequence.

11 Unscrew the four screws and the single long bolt that secure the head steady brackets to the frame and cylinder head. Removal of the bracket aids access. Slacken fully the screw clips holding the carburettors to the air filter box stubs and the inlet stubs. Slide the air hose clips to the rear so that they clear the carburettor mouths.

12 Remove the two bolts securing each inlet stub to the cylinder head. Grasp firmly the right-hand carburettor and pull it backwards, so that the air hose concertinas. This will give sufficient clearance to remove the inlet stub from the right-hand carburettor. Repeat this operation on the left-hand instrument. Pull the carburettors forwards out of the air hoses and out towards the right-hand side of the machine. Before the carburettors are completely freed for removal, the control cables must be detached. Slacken the locknuts on both throttle cable adjusters and then screw the adjusters inwards to give plenty of slack in the cable. Displace the rear cable adjuster from the abutment bracket and disconnect the inner cable from the operating pulley. Detach the forward cable in a similar manner. Slacken the choke outer clamp cable and disconnect the cable.

13 Unhook the brake light switch spring from the rear brake pedal return spring. Using a suitable lever, unhook the return spring from the brake pedal. Slacken off and remove the adjuster nut from the rear of the brake operating rod so that the rod can leave the operating arm. Push the trunnion from the arm and refit it, together with the nut, to the rod to prevent loss. Remove the brake pedal pivot bolt. Some careful manoeuvring is required if the pedal and rod are to be withdrawn without damage. Remove both front footrests, each of which is secured by a single bolt.

14 From the left-hand side of the engine, remove the gear change lever, which is secured by a circlip; and the gear change arm which is secured to the splined shaft by a pinch bolt and nut. The two components should be slid off their shafts simultaneously together with the connecting rod. The bolt must be withdrawn before the arm can be detached. Slacken evenly and then remove the bolts holding the engine left-hand cover. Pull the cover away from the engine and off the gearchange shaft. Remove the short chain protector, which is secured by a single bolt.

15 With the final drive chain spring link positioned on one of the sprockets, carefully prise off the spring link and disconnect and remove the chain.

16 The procedure for engine removal is dictated by the shape of the frame and the fact that the engine itself serves as a stressed member. Because no lower frame tubes are employed, the engine should be lowered from position, until it clears the frame, rather than lifting it out to one side as is more usual. To enable the engine to be lowered progressively in a controlled manner, a jack should be placed below the sump before mounting bolt removal. At least two people should be present during the removal operation; one to lower the jack and the other to steady the engine.

17 Remove the nuts from the four main engine mounting bolts and also from the bolts securing the engine front mounting bracket to the frame down tube. Check that all cables, leads and wires are clear of the engine and will not become snagged or trapped during engine removal. Detach the front bracket and then withdraw the engine rear mounting bolts. If necessary, take the weight of the engine to release the bolts by adjusting the jack height. With all bolts removed, the engine should be lowered steadily until the exhaust ports are level with the lower end of the front downtube. Tilt the engine backwards slightly and then lift it bodily off the jack and away from the machine, towards the right-hand side. Two people are required to lift the engine; although compact it is heavy.

5 Dismantling the engine/gearbox unit: general

1 Before commencing work on the engine unit, the external surfaces should be cleaned thoroughly. A motorcycle has very little protection from road grit and other foreign matter which

4.5 Detach securing strap and pull battery clear of carrier

4.7a Slacken the silencer/balance box joints fully

4.7b Exhaust pipes held by flange and two nuts

4.7c The single bolt retaining the silencers and pillions footrests, removed

4.7d Remove the silencers

4.7e The exhaust pipe/balance box can be removed as a unit, but separation is easier

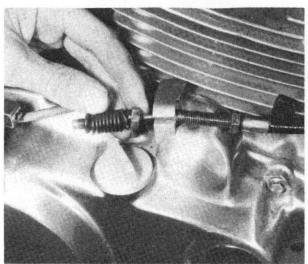
4.8a Slacken adjuster nuts fully to detach cable and ...

4.8b ... if necessary, slacken fully adjuster at handlebar

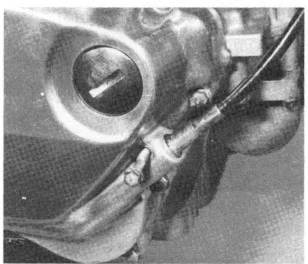
4.8c Tachometer cable is secured by a single bolt

4.9a Disconnect the neutral switch and oil pressure switch leads

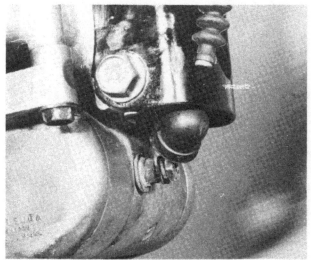
4.9b The starter cable is retained by a single nut

4.11 Detach the head steady bracket to aid carburettor removal

4.12 Detach both throttle cables after carburettor removal

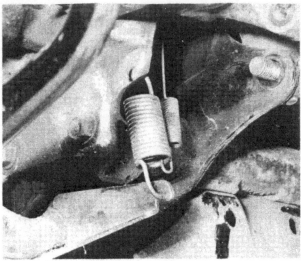
4.13a Unhook both springs and then ...

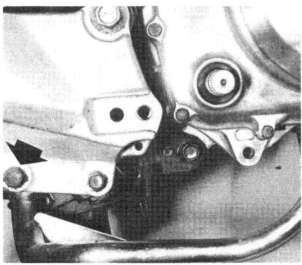
4.13b ... remove the pivot bolt to free the brake pedal (arrowed)

4.13c Remove the single bolt to detach front footrests

4.14a The gearchange arm — secured by a pinch bolt, and ...

4.14b ... the lever — held by a circlip; should be removed simultaneously

4.14c Lower chain protector is held by one bolt (arrowed)

4.15a Carefully prise off chain spring link and ...

4.15b ... run-off disconnected chain

4.17a Remove the nuts from the four engine mounting bolts on the front downtubes and ...

4.17b ... unscrew those on the upper and lower rear bolts

4.17c The engine lowered from the frame

sooner or later will find its way into the dismantled engine if this simple precaution is not carried out.

2 One of the proprietary cleaning compounds such as Gunk or Jizer can be used to good effect, especially if the compound is first allowed to penetrate the film of grease and oil before it is washed away. In the USA Gumout degreaser is an alternative.

3 It is essential when washing down to make sure that water does not enter the electrics particularly now that these parts are more vulnerable.

4 Collect together an adequate set of tools in addition to those of the tool roll carried under the seat.

5 If the engine has not been previously dismantled, an impact screwdriver will prove essential. This will safeguard the heads of the crosshead screws used for engine assembly. These are invariably machine-tightened during manufacture.

Caution: Use great care as the screws and cases are easily damaged. Use a crosshead type screwdriver and NOT one of the Phillips type, which will slip out of the screws.

6 Avoid force in any of the operations. There is generally a good reason why an item is sticking, probably due to the use of the wrong procedure or sequence of operations.

6 Dismantling the engine/gearbox: removing the cylinder head cover, rocker gear and camshaft

1 Place the engine in an upright position, supported by wooden blocks, so that it remains secure. Loosen and remove the two sleeve nuts that secure the cylinder head cover to the cylinder head. Lift the cover away.

2 Remove both sparking plugs. Apply a spanner to the alternator centre bolt and rotate the engine in a forward direction until the 'T' mark on the rotor aligns with the index mark cast into the casing. In this position both pistons will be at TDC and only one camlobe will be engaging a rocker. This will allow removal of the cylinder head without placing undue strain on the camshaft or rocker gear.

3 Slacken the cylinder head securing bolts evenly, a little at a time, following the sequence given in the accompanying illustration. Following this sequence will help prevent cylinder head distortion. Cylinder head removal should always take place when the engine is cold: this too prevents distortion.

4 Withdraw the cylinder head bolts together with the rocker pin retaining plates. Note the special brass washers fitted to the four inner bolts, which serve as oil control washers. Lift the two rocker carriers away individually, complete with the pins and rocker arms. These assemblies should be put to one side, for further attention at a later stage.

5 To enable the camshaft to be removed, the driven sprocket must first be released from the shaft flange, where it is retained by two bolts. Rotate the engine until one bolt is accessible. Slacken the bolt slightly and then rotate the engine once more, until the second bolt can be loosened and removed. Rotate the engine again and remove the first bolt. Great care should be taken not to allow any components or foreign matter to fall down through the cam chain tunnel. This is particularly important if a top-end overhaul only is to be carried out. Turn the engine, as required, until one of the cut-outs in the cam sprocket is directly above the cam journal. Slip the cam sprocket off the flange boss so that the slack gained allows disengagement of the chain. The camshaft, together with the sprocket, can now be manoeuvred from position towards the right-hand side of the machine. If a top-end overhaul only is envisaged, the cam chain should not be allowed to fall into the cam chain tunnel because retrieval is dificult. To prevent this occurring, place a short rod or screwdriver through the chain and across the cylinder head, or secure the chain with a length of stiff wire.

6.1 Loosen and remove the two head cover securing nuts

6.5a One camshaft driven sprocket retaining bolt has been removed (arrowed)

6.5b Camshaft and sprocket manoeuvred from position

7   Dismantling the engine/gearbox: removing the cylinder head, cylinder block and pistons

1   The bolts which secure the rocker gear carriers pass through the cylinder head and cylinder block and screw directly into the crankcase. It follows that once the bolts are removed, no direct restraints secure either the cylinder head or cylinder block.

2   Slacken and remove the locknut which secures the cam chain adjuster. This will allow the tensioner blade to relax completely. Remove the bolt from the rear of the cylinder head which retains the tensioner bracket. Note that this bolt, and the adjuster stud, are both fitted with a sealing O-ring to prevent oil leakage.

3   Due to the type of cylinder head gasket used, it is possible that the cylinder head will be stuck firmly to the cylinder block surface. Considerable care should be exercised when trying to separate these two components, and the temptation to use excessive force should be avoided at all costs. Commence separation by working around the head, using a block of hardwood struck by a hammer. Place the block only on portions of the cylinder head which are well supported by webs, gussets or flanges. Move around the cylinder head progressively a number of times, taking care not to damage the fins. If this method fails to loosen the two components, the application of levers will have to be adopted. Use tyre levers of reasonable length rather than screwdrivers, because they spread the load better and have no sharp edges. Insert the levers only between fins which are well supported and ensure that the areas carrying the direct load are as near as possible to the body of the cylinder/cylinder heads units. Whilst this method of cylinder head removal is not generally to be recommended, there may be no practicable alternative. It follows that extreme caution and control are required. Having accomplished separation, lift the cylinder head up off the cylinder block, preventing the cam chain from falling by placing a bar across the top of the cylinder block. The tensioner blade is secured to the top of the tensioner

arm by a clevis pin upon which it pivots. Displace the 'R' pin; withdraw the clevis pin and remove the tensioner blade. The blade must be disconnected at this stage because the clevis pin is too wide to pass down through the chain tunnel. Do not allow any of these components to fall into the chain tunnel.

4   Withdraw the cam chain guide blade from the front of the cam chain tunnel. Rotate the engine until the pistons are at TDC, hand feeding the cam chain, if necessary, to prevent it locking. Separate the cylinder block from the crankcase using a rawhide mallet or block of wood and hammer. Once again care should be taken not to damage the fins. In this case levers will not be required and should not be used. Slide the cylinder block up and off the pistons, taking care to support each piston as the cylinder block becomes free. If a top end overhaul only is being carried out, place a clean rag in each crankcase mouth before the lower edge of each cylinder frees the rings. This will preclude any small particles of broken ring falling into the crankcase. Invert the cylinder block and remove the two large O-rings, one of which fits over each protruding cylinder sleeve.

5   Prise the outer gudgeon pin circlip of each piston from position. The gudgeon pins are a light push fit in piston bosses so can be removed with ease. If any difficulty is encountered, apply to the offending piston crown a rag over which boiling water has just been poured. This will give the necessary temporary expansion to the piston bosses to allow the gudgeon pin to be pushed out. Before removing each piston, scribe the cylinder identification inside the piston skirt. A mark R or L will ensure that the piston is replaced in the correct bore, on reassembly. It is unnecessary to mark the back and front of the piston because either the inlet or exhaust valve cut-aways in the piston crown will have an 'IN' or 'EX' mark. It will be noted that on each piston the exhaust valve cut-away is slightly offset to the inside. For this reason the pistons are handed and cannot be interchanged.

6   Each piston is fitted with two compression rings and an oil control ring. It is wise to leave the rings in place on the pistons until the time comes for their examination or renewal in order to avoid confusing their correct order.

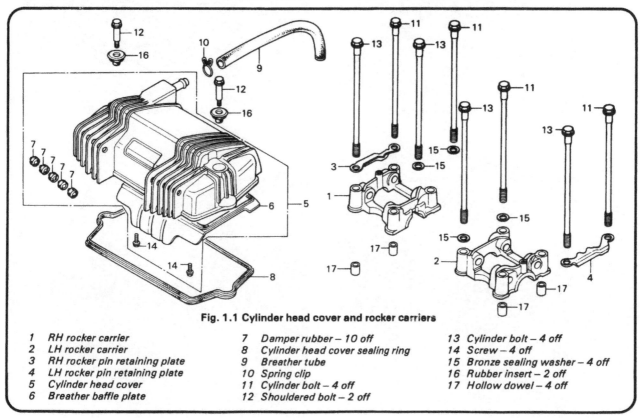

Fig. 1.1 Cylinder head cover and rocker carriers

1 RH rocker carrier	7 Damper rubber – 10 off	13 Cylinder bolt – 4 off
2 LH rocker carrier	8 Cylinder head cover sealing ring	14 Screw – 4 off
3 RH rocker pin retaining plate	9 Breather tube	15 Bronze sealing washer – 4 off
4 LH rocker pin retaining plate	10 Spring clip	16 Rubber insert – 2 off
5 Cylinder head cover	11 Cylinder bolt – 4 off	17 Hollow dowel – 4 off
6 Breather baffle plate	12 Shouldered bolt – 2 off	

7.2a Slacken and remove the locknut which secures the cam chain adjuster

7.2b Remove the bolt and adjuster stud

7.3 Lift the cylinder head up and off the cylinder block

7.4 Ease the cylinder block upwards, off the pistons

7.5a Displace the outer circlip and ...

7.5b ... push out the gudgeon pin to free the piston

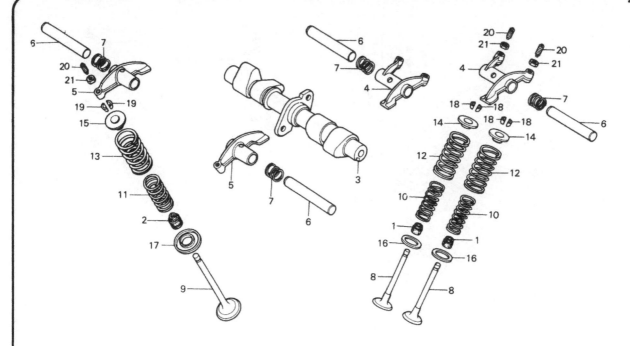

Fig. 1.2 Camshaft, valves and rocker components

1	Valve stem seal – 4 off	8	Inlet valve – 4 off
2	Valve stem seal – 2 off	9	Exhaust valve – 2 off
3	Camshaft	10	Valve inner spring – 4 off
4	Inlet valve rocker arm – 2 off	11	Valve inner spring – 2 off
5	Exhaust valve rocker arm – 2 off	12	Valve outer spring – 4 off
6	Rocker spindle – 4 off	13	Valve outer spring – 2 off
7	Spring – 4 off	14	Valve spring cap – 4 off

15	Valve spring cap – 2 off
16	Valve spring seat – 4 off
17	Valve spring seat – 2 off
18	Valve stem collet – 8 off
19	Valve stem collet – 4 off
20	Tappet adjusting screw – 6 off
21	Tappet adjusting nut – 6 off

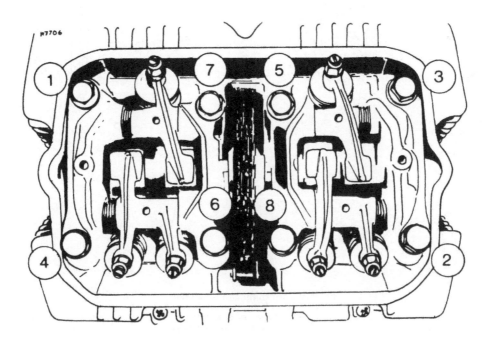

Fig. 1.3 Cylinder head bolt loosening sequence

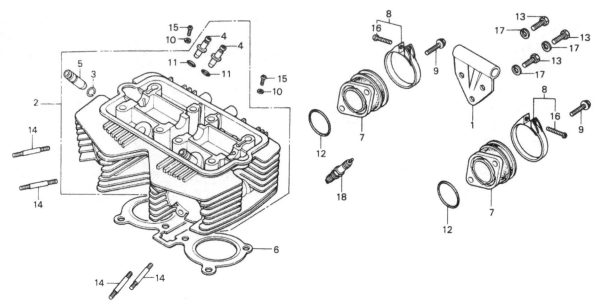

Fig. 1.4 Cylinder head

1 Head steady bracket	7 Inlet stub – 2 off	13 Bolt – 3 off
2 Cylinder head	8 Screw clip – 2 off	14 Stud – 4 off
3 Circlip – 2 off	9 Bolt – 4 off	15 Screw plug – 2 off
4 Inlet valve guide – 4 off	10 Sealing washer – 2 off	16 Screw – 2 off
5 Exhaust valve guide – 2 off	11 O-ring – 4 off	17 Spring washer – 3 off
6 Cylinder head gasket – 1 off	12 O-ring – 2 off	18 Spark plug – 2 off

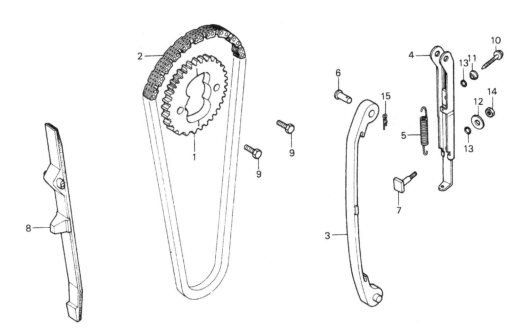

Fig. 1.5 Cam chain and tensioner

1 Camshaft sprocket	6 Chain tensioner pin	11 Washer
2 Camshaft drive chain	7 Chain tensioner adjuster	12 Sealing washer
3 Cam chain tensioner blade	8 Camshaft chain guide	13 O-ring – 2 off
4 Cam chain tensioner	9 Sprocket holding bolt – 2 off	14 Nut
5 Chain tensioner spring	10 Special bolt	15 R-pin

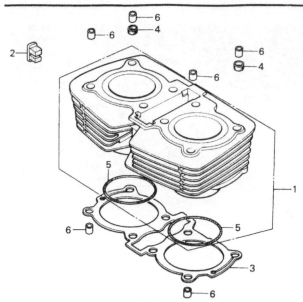

Fig. 1.6 Cylinder block

1 Cylinder block	4 Sealing ring – 2 off
2 Cable insulator	5 Cylinder base O-ring –
3 Cylinder base gasket	2 off
	6 Hollow dowel – 6 off

8 Dismantling the engine/gearbox: removing the alternator and ignition pick-up

1   To remove the alternator rotor retaining bolt it is necessary to stop the engine from rotating. This is best achieved by placing a close fitting metal bar through one or both small end eyes and allowing the bar to bear down on small wooden blocks placed across the crankcase mouth. On no account must the bar be allowed to bear directly onto the gasket face. Having stopped the engine from rotating, remove the rotor retaining bolt. Take care when loosening the bolt that the cam chain does not become trapped and that the connecting rods are at such an angle that the big-end bosses do not foul the crankcase sides.

2   Removal of the rotor bolt will give access to an internal thread in the rotor boss. This thread facilitates the use of Honda service tool No 07733 - 0020000 which should be used to

8.3a The ignition pick-up is retained by two crosshead screws

remove the rotor. If this puller is not available a legged sprocket puller **should not** be used as a substitute. Although it is possible to fit such a device it is likely to damage the alternator rotor, or the coils affixed to the back plate (stator), if special care is not taken.

3   Loosen and remove the two screws which secure the ignition pick-up to the casing and loosen the alternator lead wiring clamp screw. Pull the pick-up away from the casing, until it disengages the location dowels. Remove the three alternator stator screws, **DO NOT** loosen the two black-painted screws because this will cause loss of ignition timing accuracy. Lift the stator away together with the pick-up with which it is interconnected.

4   Prise the Woodruff key from the tapered portion of the crankshaft end and store it in a safe place. If the key is firmly located, removal is not required.

9 Dismantling the engine/gearbox: removing the gearbox final drive sprocket

1   Undo the two 6 mm bolts to release the drive sprocket fixing plate. Rotate the plate $\frac{1}{8}$th turn, then lift off the plate and the sprocket.

8.2 The rotor removed, exposing the stator and ignition pick-up

8.3b The alternator stator is held by three crosshead screws

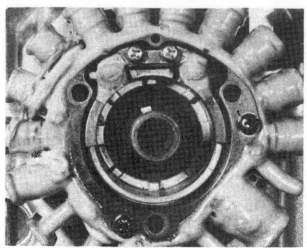

8.3c **Do not** loosen the two black-painted screws

9.1a Remove the two 6 mm bolts and ...

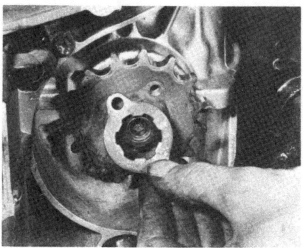

9.1b ... lift off the drive sprocket fixing plate

---

10 Dismantling the engine/gearbox: removing the primary drive cover, clutch and kickstart components

1   Remove the pinch bolt securing the kickstart lever and pull the assembly off the splined shaft. Loosen evenly and remove the bolts securing the primary drive cover to the engine. Note the clutch cable abutment bracket secured by two screws. Pull the cover from place after breaking the seal with the gasket by using a rawhide mallet.

2   Prise the rubber boot from the oil pressure switch and also the neutral indicator switch. Disconnect the electrical leads and remove them from the engine. It is recommended that the oil pressure switch be unscrewed at this juncture, because it projects and may suffer damage.

3   Unscrew the four bolts in the clutch centre evenly a little at a time in a diagonal sequence, until the pressure from the springs has been released. Remove the thrust plate and the four springs. The clutch centre is retained by a special ring nut secured by a belled washer. Removal of the nut requires a peg spanner which has four raised projections or pegs to engage with the slots in the nut. If a suitable spanner is not available, a home-made tool can be constructed easily, using a length of thick walled tubing of the correct diameter. Relieve one end of

the tube using a hacksaw and file so that the necessary projections are formed. To remove the nut, the clutch must be prevented from rotating. To accomplish this, refit temporarily the clutch springs and bolts, with a heavy washer fitted on each bolt to take the place of the clutch pressure plate. After slackening the centre nut remove the bolts, washer and springs.

4   Remove the clutch centre boss followed by the clutch plates. These may be removed individually or as a complete 'sandwich'. Note carefully the sequence of plain and friction plates, and in particular that the outermost friction plate, although generally similar to the remaining friction plates, is slightly thicker. This plate should be marked as such to avoid confusion on reassembly.

5   Remove the clutch pressure plate followed by the clutch outer drum, spacer bush and finally the heavy backing washer.

6   Grasp the outer end of the kickstart return spring with a pair of pliers. Pull the turned end from the anchor lug and allow the spring tension to be released in an anti-clockwise direction, in a controlled manner. Withdraw the spring guide and then pull the inner end of the spring from the radial hole bored in the kickstart shaft. The spring is now free to be slid off the shaft. Remove the washer, spring, kickstart pawl and the final washer from the shaft. The pawl guide plate, which is secured to the casing by two bolts need not be disturbed unless damage dictates renewal.

---

11 Dismantling the engine/gearbox: removing the oil pump and primary drive pinion

1   Release the circlip which secures the oil pump driven sprocket to the oil pump shaft. Prevent crankshaft rotation in the usual manner, using a close fitting bar passed through the small-end eye of one connecting rod. Loosen and remove the primary drive/oil pump drive sprocket centre bolt.

2   The oil pump sprockets may now be slid simultaneously off their respective shafts, still in mesh with the chain. Note and remove the shouldered spacers fitted behind the drive sprocket.

3   The oil pump is secured in place by four bolts, which should be slackened evenly, in a diagonal sequence, to prevent distortion. Remove the bolts and displace the oil pump.

4   Fitted below the oil pump is a hexagonal housing which on early models incorporates the oil system pressure release valve; on later models the valve is positioned inside the lower crankcase half. This component need not be removed unless failure to operate necessitates its examination, or evidence of oil sludging indicates the need for thorough cleaning.

5   Remove the primary drive pinion from the splined crankshaft end, and then displace the splined washer.

10.2 The oil pressure switch can be removed now

10.3a Unscrew the four clutch centre bolts evenly

10.3b A 'home-made' clutch centre nut removal tool

10.3c Fit washers to bolts and springs to secure clutch during nut removal

10.6a Pull the end of the kickstart return spring from the anchor lug (arrowed)

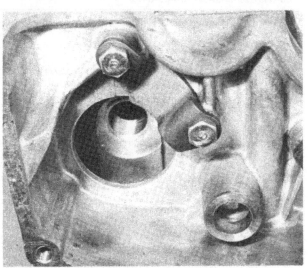

10.6b Check pawl guide plate for damage

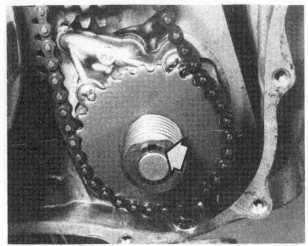

11.1a Release the circlip (arrowed) which secures the oil pump driven sprocket to the oil pump shaft

11.1b Loosen and remove centre bolt

end cover and ease the starter motor from position. If necessary, use a wooden lever between the motor front cover and the crankcase, to facilitate removal.

3   If the oil filter element was not removed before the engine was detached from the frame, this may now be accomplished. Unscrew the centre bolt which holds the filter chamber cover in place and lift it out, complete with the cover and the oil filter. The oil filter element should be pulled off the hollow bolt and discarded.

11.3 Remove the four oil pump retaining screws

12 Dismantling the engine/gearbox: removing the gearchange external components

1   Depress the pawl arm on the gearchange main arm so that the pawls clear the pins in the change drum end. With the pawl arm depressed, grasp the gearchange shaft and withdraw it from the casing.
2   Loosen and unscrew the change drum stopper arm pivot bolt until the return spring tension can be released. Unscrew the bolt fully and remove the arm.
3   Remove the change drum locating plate, which is held by countersunk screws. These screws are often very tight; take care not to damage the screw heads.
4   Unscrew the neutral indicator switch from the gearbox roof.

13 Dismantling the engine/gearbox: removing the starter motor and oil filter

1   Before continuing, the complete crankcase must be inverted on the workbench, to improve access. In order to prevent the cam chain tensioner blade from fouling the workbench, blocks must be used, placed strategically, so that the crankcase is secure.
2   Remove the two bolts which pass through the starter motor

14 Dismantling the engine/gearbox: separating the crankcase halves

1   Loosen evenly, in a diagonal sequence, and then remove, the bolts securing the two crankcase halves. The bolts should be loosened in two stages, to avoid distortion. Separate the cases using a rawhide mallet to break the seal of the gasket compound. On no account should levers of any description be placed between the mating faces of the two crankcase halves. The aluminium faces are easily damaged and this treatment will inevitably end in oil leakage at a later date.
2   Lift the lower case away, noting the position of the locating dowel pins and the figure of eight O-ring in the mating surface. All components, except the starter motor idler double gear, and kickstart assembly, will remain in the crankcase upper half.

12.4 Unscrew and remove the neutral indicator switch

15 Dismantling the engine/gearbox: removing the oil strainer unit, rear balance shaft and mainbearing support casting

1   Unscrew the two bolts which retain the oil strainer unit to the mainbearing support casting. Lift it away, noting the oil ring at the flanged end. Note that the right-hand bolt passes into the support casting in such a way as to secure the rear balance weight shaft. This will be seen more clearly when the shaft is withdrawn. To do this insert a screwdriver in the groove close to the shaft head and pull the shaft out.
2   Slacken and remove the final 6 mm bolts which pass through the support casting. One of the bolts, together with the second oil strainer bolt, secures the balance chain guide. Loosen evenly, in a diagonal sequence, the bolts clamping the support casting to the crankcase upper half; using a rawhide mallet free the casting from engagement with the hollow dowels. When the casting is free, lift it away, taking care that the lower half of each mainbearing shell does not fall out. Displaced shells should be refitted immediately in their original positions, to avoid confusion at the examination stage. Lift out the chain guide block and the rear balance weight, which will have been freed by removal of the support bracket.

16 Dismantling the engine/gearbox: removing the gearbox mainshaft and layshaft

1   Lift out the gearbox layshaft as a complete unit, putting it to one side for examination and, if necessary for further dismantling at a later stage. Similarly, remove the gearbox mainshaft.
2   Note the position of, and remove any bearing locating pins; these are a push fit in the casing. Remove the small oil feed nozzle from the mainshaft needle roller bearing housing In the crankcase half. This too is a push fit.

17 Dismantling the engine/gearbox: removing the crankshaft and front balance shaft

1   Grasp the crankshaft at both ends and lift it from position in the crankcase half. Do not displace the bearing halves at this stage. Remove the cam chain from the crankshaft drive sprocket.

2   Unscrew the balance shaft adjuster nut from the stud projecting from the primary drive chamber wall. From the opposite end of the shaft, withdraw the spring clip and displace the tensioner spring. Grasp the shaft at the adjuster end and withdraw it from the casing to free the balance weight and drive chain. Do not transpose the two balance weights or shafts.

18 Dismantling the engine/gearbox: removing the gear selector mechanism

1   Slide out the rear selector fork rod and lift out the two selector forks. Similarly, remove the front rod and the single selector rod. Replace the forks on their respective rods in the correct position, to aid reassembly.
2   The gear change drum is now free to be slid out of the casing, towards the primary drive side of the engine.
3   Take care not to lose the selector fork guide pins. These are a light push fit in the forks and consequently fall out easily.

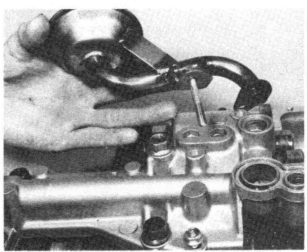

15.1 Lift away the oil strainer unit with the two bolts removed

17.2a Unscrew the balance shaft adjuster nut and ...

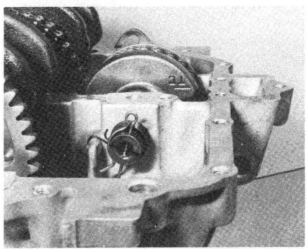

17.2b ... at the other end, withdraw the spring clip and displace the spring

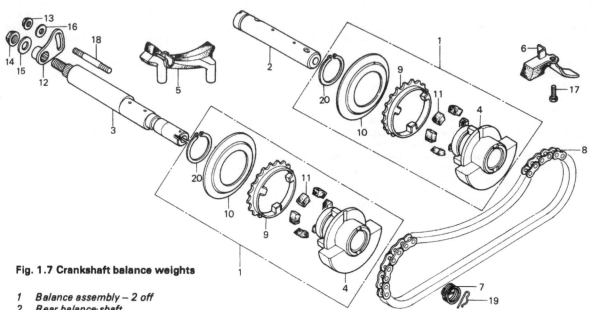

Fig. 1.7 Crankshaft balance weights

1   Balance assembly – 2 off
2   Rear balance shaft
3   Front balance shaft
4   Balance weight – 2 off
5   Chain tensioner
6   Chain guard plate
7   Return spring
8   Chain
9   Balance weight sprocket – 2 off
10 Balance plate – 2 off

11 Damper rubber – 12 off
12 Chain tension adjuster quadrant
13 Nut
14 Nut
15 Washer

16 Washer
17 Bolt
18 Stud bolt
19 R-pin
20 Circlip – 2 off

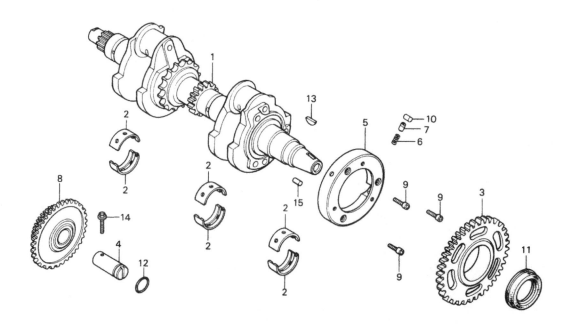

Fig. 1.8 Crankshaft assembly

1   Crankshaft assembly
2   Main bearing shell – 6 off
3   Starter driven pinion
4   Starter idler gear shaft
5   Starter clutch

6   Spring – 3 off
7   Plunger – 3 off
8   Starter idler gear
9   Bolt – 3 off

10 Roller – 3 off
11 Oil seal
12 O-ring
13 Woodruff key

19 Dismantling the engine/gearbox: removing the starter motor idler gear and kickstart assembly

1   Removal of these items is not required unless wear is suspected or damage has occurred, requiring renewal.

2   The starter motor idler gear runs on a plain shaft which is a push fit in the casing. The shaft is secured by a single bolt. Remove the bolt, withdraw the shaft and lift out the idler gear. Note the O-ring fitted to the annular groove in the shaft.

3   Displace the 'E' clip from the inner end of the kickstart shaft (where fitted). Withdraw the shaft and displace the kickstart pinion. Note the washer that lies between the pinion and the support lug in the casing.

4   On later models with the oil pressure relief valve positioned inside the crankcase, note that this can now be removed for inspection or renewal.

20 Examination and renovation: general

1   Before examining the component parts of the dismantled engine/gearbox unit for wear, it is essential that they should be cleaned thoroughly. Use a paraffin/petrol mix to remove all traces of oil and sludge which may have accumulated within the engine.

2   Examine the crankcase castings for cracks or other signs of damage. If a crack is discovered, it will require professional attention or in an extreme case, renewal of the casting.

3   Examine carefully each part to determine the extent of wear. If in doubt, check with the tolerance figures whenever they are quoted in the text. The following sections will indicate what type of wear can be expected and in many cases, the acceptable limits.

4   Use clean, lint-free rags for cleaning and drying the various components, otherwise there is risk of small particles obstructing the internal oilways.

21 Examination and renovation: big-end bearings

1   Big-end failure is invariably indicated by a pronounced knock from the crankcase. The knock will become progressively worse, accompanied by vibration. It is essential that the bearings are renewed as soon as possible since the oil pressure will be reduced and damage caused to other parts of the engine.

2   The big-ends have shell type bearings. Examine the bearing surfaces after removing the bearing caps; it is not necessary to remove the shell. If the bearings are badly scuffed or scored the shells will have to be renewed. Always renew bearings as a complete set. If the shell surfaces are excessively scuffed or have 'picked-up' and the journal on that bearing is 'blued' or discoloured, it may be an indication of lubrication failure. The functioning of the lubrication system MUST be checked before engine reassembly in such a case.

3   If the condition of the bearings appears to be satisfactory, check that the clearance between each bearing and the journal is within the recommended limit as laid down in the specifications at the beginning of this Chapter. The clearance can be assessed by measuring the internal diameter of the shell bearing and the outside diameter of the big-end journal and subtracting the second figure obtained from the first. Note that if measuring the bearing in this way, care must be taken not to damage the shell's soft material with the measuring equipment. 'Plastigauge' can also be used in the following manner. Cut a short length of 'Plastigauge' and place it on the journal so that the pre-marked indexing runs axially. Bolt up the connecting rod and bearing cap to the recommended torque of 18 - 21 lb ft (250 - 290 kg cm). Do not rotate the bearing. Separate the two bearing halves and remove and inspect the 'Plastigauge'. This will indicate the amount of clearance; reading of the 'Plastigauge' will depend on the type used. See manufacturer's instructions. The clearance may also be ascertained by direct measurement of the thickness of the 'Plastigauge', using a micrometer.

4   When renewing the bearings, bearing selection should be made by referring to the connecting rod code, a number stamped on the machined side of the connecting rod, and the big-end journal code, which takes the form of letters scribed on the adjacent crankshaft webs. Refer to the accompanying illustrations for the positions of the various code numbers on the crankshaft. The letters and numbers should be cross-referred to the accompanying selection chart.

5   When fitting new bearings, ensure that they are positioned correctly and that the tongues on the end of each shell locate with the recesses in the connecting rod or bearing cap. Also check the clearance on each bearing to ensure that selection is accurate. It is considered good practice to renew all bearing shells when an engine is stripped down, irrespective of their condition. Shell bearings are relatively inexpensive compared with the cost of subsequent dismantling to replace bearings that have failed prematurely.

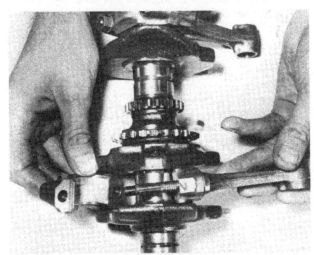

21.2a Remove the cap nuts and separate the connecting rod and cap

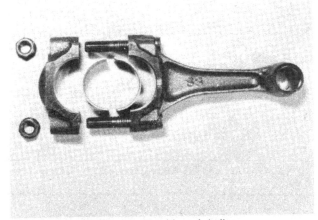

21.2b The connecting rod assembly and shells

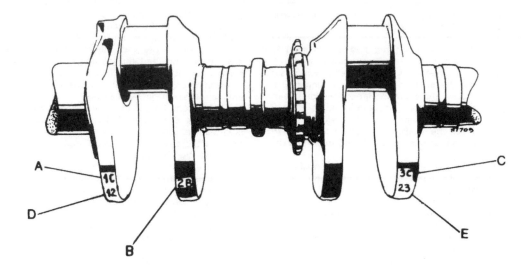

Fig. 1.9 Main bearing and big-end journal code locations

A = LH main bearing journal (code letter C)
B = Centre main bearing journal (code letter B)
C = RH main bearing journal (code letter C)

D = LH big-end bearing journal (code number 2)
E = RH big-end bearing journal (code number 3)

CB250 MODELS			CRANKPIN O.D. CODE NUMBERS		
			1	2	3
			29.992 ~ 30.000 mm	29.984 ~ 29.992 mm	29.976 ~ 29.984 mm
CONNECTING ROD I.D. CODE NUMBERS	1	33.000 ~ 33.008 mm	E (YELLOW)	D (GREEN)	C (BROWN)
	2	33.008 ~ 33.016 mm	D (GREEN)	C (BROWN)	B (BLACK)
	3	33.016 ~ 33.024 mm	C (BROWN)	B (BLACK)	A (BLUE)

CB400 MODELS			CRANKPIN O.D. CODE NUMBERS		
			1	2	3
			35.992 ~ 36.000 mm	35.984 ~ 35.992 mm	35.976 ~ 35.984 mm
CONNECTING ROD I.D. CODE NUMBERS	1	39.000 ~ 39.008 mm	E (YELLOW)	D (GREEN)	C (BROWN)
	2	39.008 ~ 39.016 mm	D (GREEN)	C (BROWN)	B (BLACK)
	3	39.016 ~ 39.024 mm	C (BROWN)	B (BLACK)	A (BLUE)

Bearing insert thickness:
A (Blue)   – 1.502 to 1.506 mm
B (Black)   – 1.498 to 1.502 mm
C (Brown)   – 1.494 to 1.498 mm
D (Green)   – 1.490 to 1.494 mm
E (Yellow) – 1.486 to 1.490 mm

Fig. 1.10 Big-end bearing selection chart

22 Examination and renovation: crankshaft and main bearings

1   Examine the main bearing shells and check the clearances, using the procedure as described for big-end bearings. The correct clearances are as shown in the specifications.
2   Measure the main bearing and big-end bearing journals for ovality by measuring each journal in several radial positions, with a micrometer. If any journal is worn beyond the service limit, the crankshaft must be renewed.

3   Support the crankshaft at each end on vee-blocks or between centres. Rotate the crankshaft and measure the runout at the centre main bearing journal, using a dial gauge. If the runout exceeds the service limit, the crankshaft must be renewed, or placed in the hands of a specialist for straightening. Bear in mind that the runout is half the actual reading taken.
4   Main bearing shells must be selected by referring to the code numbers scribed on the adjacent crankshaft webs and by the main bearing cap code numbers or letters which are marked on the crankcase. The coding on the crankcase can be found stamped on the outside of the gearbox rear wall. Cross refer the main bearing journal letters and the cap letters or numbers with the accompanying chart for correct bearing selection.
5   Check the security of the ball bearings which are used to plug the oilways in the crankshaft. They occasionally work loose, causing lubrication problems and subsequent bearing failure. A loose ball can be carefully caulked back into place, after being secured by the application of a locking fluid.

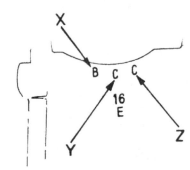

Fig. 1.11 Bearing holder diameter code number location on outer rear wall of gearbox

X = LH main bearing housing code letter
Y = Centre main bearing housing code letter
Z = RH main bearing housing code letter

CB250 MODELS			MAIN JOURNAL O.D. CODES		
			A	B	C
			29.992 ~ 30.000 mm	29.984 ~ 29.992 mm	29.976 ~ 29.984 mm
CASE I.D. CODES	A	33.000 ~ 33.008 mm	E (YELLOW)	D (GREEN)	C (BROWN)
	B	33.008 ~ 33.016 mm	D (GREEN)	C (BROWN)	B (BLACK)
	C	33.016 ~ 33.025 mm	C (BROWN)	B (BLACK)	A (BLUE)

CB400 MODELS			MAIN JOURNAL O.D. CODES		
			A	B	C
			35.992 ~ 36.000 mm	35.984 ~ 35.992 mm	35.976 ~ 35.984 mm
CASE I.D. CODES	A	39.000 ~ 39.008 mm	E (YELLOW)	D (GREEN)	C (BROWN)
	B	39.008 ~ 39.016 mm	D (GREEN)	C (BROWN)	B (BLACK)
	C	39.016 ~ 39.025 mm	C (BROWN)	B (BLACK)	A (BLUE)

Bearing insert thickness:
A (Blue)   – 1.502 to 1.506 mm
B (Black)   – 1.498 to 1.502 mm
C (Brown)   – 1.494 to 1.498 mm
D (Green)   – 1.490 to 1.494 mm
E (Yellow) – 1.486 to 1.490 mm

Fig. 1.12 Main bearing selection table

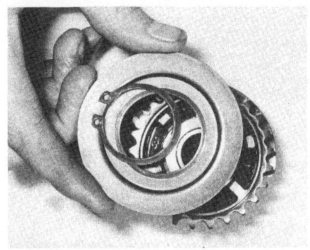

24.2 Balance weights incorporate simple shock absorbers

23 Examination and renovation: connecting rods

1   It is unlikely that either of the connecting rods will bend during normal usage, unless an unusual occurrence such as a dropped valve has caused the engine to lock. Carelessness when removing a tight gudgeon pin can also give a rise to a similar problem. It is not advisable to straighten a bent connecting rod; renewal is the only satisfactory solution.
2   The small-end eye of the connecting rod is unbushed and it will be necessary to renew the connecting rod if the gudgeon pin becomes a slack fit. Always check that the oil hole in the small-end eye is not blocked since if the oil supply is cut off, the bearing surfaces will wear very rapidly.

24 Examination and renovation: balance weights and drive chain

1   When checking the balance weight assemblies it is best to check each set independently, to avoid accidental interchange of the component parts.
2   Displace the circlip from the balance weight centre boss and remove the side plate. Before removing the sprocket and damper rubber inserts note that the sprocket is timed in relation to the balance weight. The punch mark on each should be in alignment, to ensure that they are in the correct relative position.
3   Measure each shaft and its corresponding bearing diameter with a micrometer or vernier gauge. The correct sizes are as follows:

Balance shaft OD (min)   17.95 mm (0.707 in)
Balance shaft ID (max)   18.04 mm (0.710 in)

4   Damage to compression of the damper rubbers will be obvious on inspection. If necessary, the rubbers should be renewed as a complete set.
5   When reassembling the balance weight assembly ensure that the alignment marks are correctly placed.
6   The balance weight drive chain operates in almost ideal conditions, being lubricated with a constant supply of filtered oil, and therefore does not normally wear until a considerable mileage has been covered. If damage to the chain links or rollers is evident, or if play in the chain is excessive, it must be renewed.

25 Examination and replacement: oil seals

1   An oil seal is fitted to the right-hand end of the crankshaft assembly, to prevent oil from entering the contact breaker. There is also an oil seal fitted behind the gearbox final drive sprocket. If either seal is damaged or has shown a tendency to leak, it must be renewed.
2   Oil seals tend to lose their effectiveness if they harden with age. It is difficult to give any firm recommendations in this respect except to say that if there is any doubt about the condition of a seal, renew it as a precaution.

26 Examination and renovation: cylinder block

1   The usual indication of badly worn cylinder bores and pistons is excessive smoking from the exhausts, high crankcase compression which causes oil leaks, and piston slap, a metallic rattle that occurs when there is little or no load on the engine. If the top of the cylinder bore is examined carefully, it will be found that there is a ridge at the front and back the depth of which will indicate the amount of wear which has taken place. This ridge marks the limit of travel of the top piston ring.
2   Since there is a difference in cylinder wear in different directions, side to side and back to front measurements should be made. Take measurements at three different points down the length of the cylinder bore, starting just below the top piston ring ridge, then about $2\frac{1}{2}$ inch below the top of the bore and the last measurement about 1 inch from the bottom of the cylinder bore. Refer to the Specifications for cylinder specifications. If any of the cylinder bore inside diameter measurements exceed the service limit the cylinder must be bored out to take the next size of piston. If there is a difference of more than 0.004 in between any two measurements the cylinder should, in any case, be rebored.
3   Oversize pistons are available in four sizes, + 0.25 mm (0.010 in), + 0.50 mm (0.020 in), + 0.75 mm (0.030 in) and + 1.0 mm (0.040 in).
4   Check that the surface of the cylinder bore is free from score marks or other damage that may have resulted from an earlier engine seizure or a displaced gudgeon pin. A rebore will be necessary to remove any deep scores, irrespective of the amount of bore wear that has taken place, otherwise a compression leak will occur.
5   Make sure the external cooling fins of the cylinder block are not clogged with oil or road dirt which will prevent the free flow of air and cause the engine to overheat.

27 Examination and renovation: pistons and piston rings

1   Attention to the pistons and piston rings can be overlooked if a rebore is necessary, since new components will be fitted.

2   If a rebore is not necessary, examine each piston carefully. Reject pistons that are scored or badly discoloured as the result of exhaust gases by-passing the rings. Remove the piston rings by pushing the ends apart with the thumbs whilst gently easing the ring from its groove. Great care is necessary throughout this operation because the rings are brittle and will break easily, if overstressed. If the rings are gummed in the grooves, three strips of tin can be used to ease them free, as shown in the accompanying illustration.

3   Remove all carbon from the piston crowns, using a blunt scraper, which will not damage the surface of the piston. Clean away carbon deposits from the valve cutaways and finish off with metal polish so that a smooth, shining surface is achieved. Carbon will not adhere so readily to a polished surface.

4   Small high spots on the back and front areas of the piston can be carefully eased back with a fine swiss file. Dipping the file in methylated spirit or rubbing its teeth with chalk will prevent the file clogging and eventually scoring the piston. Only very small quantities of material should be removed, and never enough to interfere with the correct tolerances. Never use emery paper or cloth to clean the piston skirt; the fine particles of emery are inclined to embed themselves in the soft aluminium and consequently accelerate the rate of wear between bore and piston.

5   Measure the outside diameter of the piston about 0.2 in up from the skirt at right angles to the line of the gudgeon pin. If the measurement is under the service limit the piston should be renewed.

6   Check that the gudgeon pin bosses are not worn or the circlip grooves damaged. Check that the piston ring grooves are not enlarged. Side float should not exceed the recommended amount.

7   Piston ring wear can be measured by inserting the rings in the bore from the top and pushing them down with the base of the piston so that they are square with the bore and close to the bottom of the bore where the cylinder wear is least. Place a feeler gauge between the ring ends. If the clearance exceeds the service limit the ring should be renewed.

8   Check that there is no build up of carbon either in the ring grooves or the inner surfaces of the rings. Any carbon deposits should be carefully scraped away. A short length of old piston ring fitted with a handle and sharpened at one end to a chisel point is ideal for scraping out encrusted piston ring grooves.

9   All pistons have their size stamped on the piston crown, original pistons being stamped standard (STD) and oversize pistons having the amount of oversize indicated.

27.3 Cleaned pistons should have a smooth shining surface

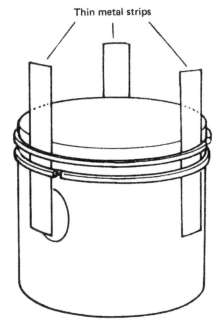

Fig. 1.14 Freeing gummed rings

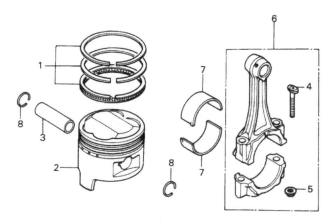

Fig. 1.13 Pistons and connecting rods

1   Piston ring set – 2 off
2   Piston – 2 off
3   Gudgeon pin – 2 off
4   Bolt – 4 off
5   Nut – 4 off
6   Connecting rod assembly – 2 off
7   Big-end bearing shell – 4 off
8   Circlip – 4 off

28 Examination and renovation: cylinder head and valves

1 It is best to remove all carbon deposits from the combustion chambers before removing the valves for inspection and grinding-in. Use a blunt end chisel or scraper so that the surfaces are not damaged. Finish off with a metal polish to achieve a smooth, shining surface. If a mirror finish is required, a high speed felt mop and polishing soap may be used. A chuck attached to a flexible drive will facilitate the polishing operation.

2 A valve spring compression tool must be used to compress each set of valve springs in turn, thereby allowing the split collets to be removed from the valve cap and the valve springs and caps to be freed. Keep each set of parts separate and mark each valve so that it can be replaced in the correct combustion chamber. There is no danger of inadvertently replacing an inlet valve in an exhaust position, or vice-versa, as the valve heads are of different sizes. The normal method of marking valves for later identification is by centre punching them on the valve head. This method is not recommended on valves, or any other highly stressed components, as it will produce high stress points and may lead to early failure. Tie-on labels, suitably inscribed, are ideal for the purpose.

3 Before giving the valves and valve seats further attention, check the clearance between each valve stem and the guide in which it operates. Measure also the valve stem diameters. Measure the valve stem at the point of greatest wear and then measure again at right-angles to the first measurement. If the valve stem is below the service limit, it must be renewed. The valve stem/guide clearance can be measured with the use of a dial gauge and a new valve. Place the new valve into the guide and measure the amount of shake with the dial gauge tip resting against the top of the stem. If the amount of wear is greater than the wear limit, the guide must be renewed.

4 To remove an old valve guide, place the cylinder head in an oven and heat it to about 150°C. The old guide can now be tapped out from the cylinder side. The correct drift should be shouldered with the smaller diameter the same size as the valve stem and the larger diameter slightly smaller than the OD of the valve guide. If a suitable drift is not available a plain brass drift may be utilised with great care. **DO NOT** use a blowtorch or other localised form of heat source to heat the cylinder head prior to valve guide removal. Heating in this way may cause irreparable damage due to warping. Each inlet valve guide is fitted with an O-ring to ensure perfect sealing. The O-rings must be renewed with new components. The exhaust valve guides are fitted with a circlip which acts as a positive stop in positioning the guide in the housing. The circlips should be renewed at the same time as the guides. New valve guides should be fitted at the same heat as for removal.

5 Valve grinding is a simple task. Commence by smearing a trace of fine valve grinding compound (carborundum paste) on the valve seat and apply a suction tool to the head of the valve. Oil the valve stem and insert the valve in the guide so that the two surfaces to be ground in make contact with one another. With a semi-rotary motion, grind in the valve head to the seat, using a backward and forward action. Lift the valve occasionally so that the grinding compound is distributed evenly. Repeat the application until an unbroken ring of light grey matt finish is obtained on both valve and seat. This denotes the grinding operation is now complete. Before passing to the next valve, make sure that all traces of the valve grinding compound have been removed from both the valve and its seat and that none has entered the valve guide. If this precaution is not observed, rapid wear will take place due to the highly abrasive nature of the carborundum paste.

6 When deep pits are encountered, it will be necessary to use a valve refacing machine and a valve seat cutter, set to an angle of 45°. Never resort to excessive grinding because this will only pocket the valves in the head and lead to reduced engine efficiency. If there is any doubt about the condition of a valve, fit a new one.

7 Examine the condition of the valve collets and the groove on the valve stem in which they seat. If there is any sign of damage, new parts should be fitted. Check that the valve spring collar is not cracked. If the collets work loose or the collar splits whilst the engine is running, a valve could drop into the cylinder and cause excessive damage.

8 Check the free length of each of the valve springs. The springs have reached their serviceable limit when they have compressed to the limit readings given in the Specifications section of this Chapter.

9 Reassemble the valve and valve springs by reversing the dismantling procedure. Fit new oil seals to each valve guide and oil both the valve stem and the valve guide, prior to reassembly. Take special care to ensure the valve guide oil seal is not damaged when the valve is inserted. The springs are of variable pitch. Each must be fitted with the closer coils nearest the valve guide. As a final check after assembly, give the end of each valve stem a light tap with a hammer, to make sure the split collets have located correctly.

10 Check the cylinder head for straightness, especially if it has shown a tendency to leak oil at the cylinder head joint. If there is any evidence of warpage, provided it is not too great, the cylinder head must be either machined flat or a new head fitted. Most cases of cylinder head warpage can be traced to unequal tensioning of the cylinder head bolts by tightening them in incorrect sequence.

28.9a Note the spring seat fitted below oil seal

28.9b Lubricate the valve stem before insertion

28.9c Springs must be fitted with close coils downwards

28.9d Compress springs to allow collet reinstallation

29 Examination and renovation: rocker spindles and rocker arms

1   The rocker carriers together with their related components should be considered as two separate assemblies and dismantled separately, to prevent the accidental interchange of components. The various parts of each assembly should be kept in matched sub-assemblies for the same reason.

2   Displace each rocker pin so that the rocker arms and the end float control springs can be removed. Take care that each spindle remains with its original rocker arm.

3   Check the rocker arms for undue wear on their spindles and renew any that show excessive play. Examine each rocker arm where it bears on the cam and the adjuster end which bears on the valve stem head. Arms that are badly hammered or worn should be renewed. Slight wear marks may be stoned out with a carborundum oil stone, but remember that if too much metal is removed it will not only weaken the component but may make correct adjustment of the tappet gap impossible.

30 Examination and renovation: camshaft, camshaft bearings, drive chain and tensioner

1   The camshaft should be examined visually for wear, which will probably be most evident on the ramps of each cam and where the cam contour changes sharply. Also check the bearing surfaces for obvious wear and scoring. Cam lift can be checked by measuring the height of the cam from the bottom of the base circle to the top of the lobe. If the measurement is less than the service limit the opening of that particular valve will be reduced, resulting in poor performance. Measure the diameter of each bearing journal with a micrometer or vernier gauge. If the diameter is less than the service limit, renew the camshaft.

2   The camshaft bears directly on the cylinder head material and that of the rocker carriers, there being no separate bearings. Check the bearing surfaces for wear and scoring. The clearance between the camshaft bearing journals and the bearing surfaces may be checked using Plastigauge (press gauge) material in the same manner as described for crankshaft bearing clearance in Section 22.3 of this Chapter. If the clearance is greater than given for the service limit, the recommended course is to renew the camshaft. If bad scuffing is evident on the camshaft bearing surfaces of the cylinder head and carriers due to a lubrication failure, the only remedy is to renew the cylinder head rocker carriers, and the camshaft if it transpires that it has been damaged also.

3   Check the camshaft drive chain for wear and damage to the side plates and pins. No service data is available for this chain, but in common with the balance weight chain, working conditions are ideal and therefore a low wear rate may be expected. It should, however, be noted that the breakage of any chain within the engine will not only necessitate complete engine dismantling for renewal but may cause extensive engine damage.

4   Check the cam chain tensioner blade and the guide blade for wear and for separation of the rubber from the backing piece. No specifications are laid down for acceptable blade wear, but it is suggested that the components are renewed if wear has reduced the thickness of rubber to more than 50%. If the rubber has begun to separate from the blade, the component should be renewed as a matter of course.

31 Examination and renovation: camshaft drive chain sprockets

1   The upper camshaft chain sprocket is bolted to the camshaft and in consequence is easily renewable if the teeth become hooked, worn, chipped or broken. The lower sprocket is integral with the crankshaft and if any of these defects are evident, the complete crankshaft assembly must be renewed. Fortunately, this drastic course of action is rarely necessary since the parts concerned are fully enclosed and well lubricated, working under ideal conditions.

2   If the sprockets are renewed, the chain should be renewed at the same time. It is bad practice to run old and new parts together since the rate of wear will be accelerated.

32 Examination and renovation: starter motor clutch and drive pinion

1   The starter motor clutch is a one-way clutch working on the roller and ramp principle and is fitted to the left-hand end of the crankshaft. In the normal course of events, it is unlikely that the clutch will malfunction.

2   To check whether the clutch is functioning correctly, grasp the crankshaft and rotate the sprocket. When rotated in an anti-clockwise direction, as viewed from the sprocket side, the clutch should lock immediately, allowing power to be transmitted from the sprocket to the crankshaft. When rotated clockwise, the sprocket should be free to run smoothly. If the movement is unsatisfactory, withdraw the sprocket from the clutch and pull it off the crankshaft. The sprocket boss should be smooth; only scoring indicates that the rollers are similarly marked and require further inspection.

3   To gain access to the rollers, springs and plungers, removal of the clutch body from the crankshaft is required. The clutch body is retained by three socket screws of a particular type that require a specially shaped socket key for their removal. Due to the unusual screw form, the availability of the socket is restricted and therefore it is recommended that a Honda Service Agent be approached to carry out removal and replacement.

4   After removing the screws, pull the clutch body off the crankshaft. Displace the rollers, springs and plungers. If there is any obvious wear or damage to these components, replacements should be fitted.

33 Examination and renovation: clutch assembly

1   After an extended period of service the clutch linings will wear and promote clutch slip. Measure the width of each friction plate and compare the results with the specifications given. Note that the first friction plate to be removed during dismantling (the outermost plate) is slightly thicker than the remainder of the plates. This plate should be marked appropriately, to avoid confusion. When the overall width reaches the limit, the inserted plates must be renewed, preferably as a complete set.

2   The plain plates should not show any excess heating (blueing). Check the warpage of each plain plate using plate glass or a surface plate and a feeler gauge. The maximum allowable warpage is 0.20 mm (0.008 in).

3   The plain backing plate is fitted to the clutch centre is retained by a double spiral circlip. Fitted behind the plate is a cone spring which acts as a shock absorber. To free the plate, spring and spring seat the circlip must be prised from position. Start at the outer end of the clip displacing it from the groove using a small screwdriver, and work round the clutch centre until the clip is clear of the groove. Before removing these items check the gap between the plain backing plate and the clutch centre flange. For the shock absorber to function efficiently there must be a gap of at least 0.10 mm (0.004 in).

4   Check the free length of each clutch spring with a vernier gauge. After considerable use, the springs will take a permanent set, thereby reducing the pressure applied to the clutch plates. The correct measurements are given in the Specifications section of this Chapter.

5   Examine the clutch assembly for burrs or indentation on the edges of the protruding tongues of the inserted plates and/or slots worn in the edges of the outer drum with which they engage. Similar wear can occur between the inner tongues of the plain clutch plates and the slots in the clutch inner drum. Wear of this nature will cause clutch drag and slow disengage-ment during gear changes, since the plates will become trapped and will not free fully when the clutch is withdrawn. A small amount of wear can be corrected by dressing with a fine file; more extensive wear will necessitate renewal of the worn parts.

6   The clutch release mechanism attached to the primary drive cover does not normally require attention because it is well lubricated. In due course the ball bearing may wear due to the thrust imposed during clutch operation. The bearing is a push fit in the thrust plate and is easily removed for renewal.

34 Examination and renovation: gearbox components

1   Examine each of the gear pinions to ensure that there are no chipped or broken teeth and that the dogs on the end of the pinions are not rounded. Gear pinions with any of these defects must be renewed; there is no satisfactory method of reclaiming them.

2   The gearbox bearings must be free from play and show no signs of roughness when they are rotated. After thorough washing in petrol, the bearings should be examined for roughness and play. Also check for pitting on the roller tracks.

3   It should not be necessary to dismantle the gear cluster unless damage has occurred to any of the pinions or a fault has become apparent in the gearbox shaft.

4   The accompanying illustrations show how the clusters are arranged on their shafts. It is imperative that the gear clusters, including the thrust washers and circlips, are assembled in **EXACTLY** the correct sequence, otherwise constant gear selection problems will arise.

5   In order to eliminate the risk of incorrect reassembly, make a rough sketch as the pinions are removed. Also strip and rebuild as soon as possible, to reduce any confusion which might occur at a later date.

6   It is advisable to renew the gearbox oil seals irrespective of their condition. Should a re-used oil seal fail at a later date, a considerable amount of work is involved to gain access to renew it.

7   Check the gear selector rods for straightness by rolling them on a sheet of plate glass. A bent rod will cause difficulty in selecting gears and will make the gear change particularly heavy.

8   The selector forks should be examined closely, to ensure that they are not bent or badly worn. The case-hardened pegs which engage with the cam channels are easily renewable if they are worn. Under normal circumstances, the gear selector mechanism is unlikely to wear quickly, unless the gearbox oil level has been allowed to become low.

9   The tracks in the selector drum, with which the selector forks engage, should not show any undue signs of wear unless neglect has led to under-lubrication of the gearbox. Check the tension of the gearchange pawl, gearchange arm and drum stopper arm springs. Weakness in the springs will lead to imprecise gear selection. Check the condition of the gear stopper arm roller and the pins in the change drum end with which it engages. It is unlikely that wear will take place here except after considerable mileage.

35 Examination and renovation: kickstart components – CB250 N and CB400 N only

1   Because the kickstart assembly must be dismantled completely as an integral part of removal from the casing, no further dismantling is required for inspection.

2   If slipping has occurred, check the kickstart wheel and ratchet piece for wear. Damage will be self-evident. Inspect also the pawl or ratchet piece and the stop plate for indications of imminent failure. The kickstart return spring should be examined closely for signs of fatigue or overstraining. It should be noted that renewal of the kickstart spring can be made without separating the crankcase.

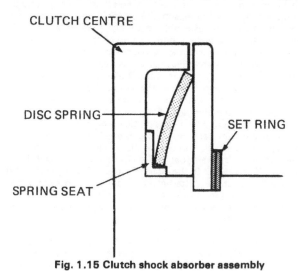

CLUTCH CENTRE

DISC SPRING

SET RING

SPRING SEAT

Fig. 1.15 Clutch shock absorber assembly

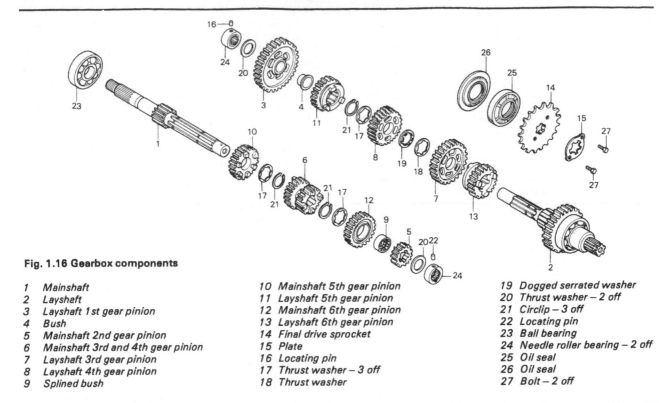

Fig. 1.16 Gearbox components

1 Mainshaft	10 Mainshaft 5th gear pinion	19 Dogged serrated washer
2 Layshaft	11 Layshaft 5th gear pinion	20 Thrust washer – 2 off
3 Layshaft 1st gear pinion	12 Mainshaft 6th gear pinion	21 Circlip – 3 off
4 Bush	13 Layshaft 6th gear pinion	22 Locating pin
5 Mainshaft 2nd gear pinion	14 Final drive sprocket	23 Ball bearing
6 Mainshaft 3rd and 4th gear pinion	15 Plate	24 Needle roller bearing – 2 off
7 Layshaft 3rd gear pinion	16 Locating pin	25 Oil seal
8 Layshaft 4th gear pinion	17 Thrust washer – 3 off	26 Oil seal
9 Splined bush	18 Thrust washer	27 Bolt – 2 off

36 Examination and renovation: primary drive gears

1   Examine the primary drive pinions for chipped, broken or worn teeth. This type of damage cannot be rectified; the component must be renewed. The pinions concerned are unlikely to become damaged unless a stray engine component becomes trapped between the teeth. The primary driven gear is riveted to the clutch outer drum. These two items are not supplied separately as replacements. If one fails the other must be renewed at the same time.

37 Engine reassembly: general

1   Before reassembly of the engine/gearbox unit is commenced, the various component parts should be cleaned thoroughly and placed on a sheet of clean paper, close to the working area.
2   Make sure all traces of old gaskets have been removed and that the mating surfaces are clean and undamaged. One of the best ways to remove old gasket cement is to apply a rag soaked in methylated spirit. This acts as a solvent and will ensure that the cement is removed without resort to scraping and the consequent risk of damage.
3   Gather together all the necessary tools and have available an oil can filled with clean engine oil. Make sure all new gaskets and oil seals are to hand, also all replacement parts required. Nothing is more frustrating than having to stop in the middle of a reassembly sequence because a vital gasket or replacement has been overlooked.
4   Make sure that the reassembly area is clean and that there is adequate working space. Refer to the torque and clearance settings wherever they are given. Many of the smaller bolts are easily sheared if over-tightened. Always use the correct size screwdriver bit for the crosshead screws and never an ordinary screwdriver or punch. If the existing screws show evidence of maltreatment in the past, it is advisable to renew them as a complete set.

38 Engine reassembly: replacing the crankshaft assembly and front balance weight

1   Place the upper crankcase half so that it rests securely, upside down, supported on blocks at such a height that when fitted, the cam chain tensioner blade will not foul the workbench.
2   Mesh the drive chain with the front balance weight, and with the sprocket towards the primary drive side of the engine, install the assembly in the casing. Lubricate the balance shaft and slide it into position in the casing so that the adjuster quadrant elongated hole fits over the adjuster stud. With the outer turned end innermost, slide the tensioner spring onto the shaft so that the outer turned end engages with the casing lug. Tension the spring in an anti-clockwise direction, until the inner end can be slid into the slotted shaft end. The spring should be tensioned through about ¾ of a turn, and then secured using the spring clip. If, for some reason, the adjuster piece was removed from the shaft, the assembly must be fitted in a particular sequence. Install the shaft in the casing so that the punch mark on the threaded end is in the 10 o'clock position. Fit the spring and then rotate the shaft clockwise - as seen from the threaded end - until the punch mark is in the 6 o'clock position. Hold the shaft in this position, against the spring tension, and fit the adjuster piece to the splined shaft end so that the adjuster stud lies centrally in the elongated hole. Fit and tighten the retaining nut. Secure the spring by means of the spring clip.
3   Lubricate with engine oil the main bearing shells, after ensuring that they are correctly in place. Fit the camshaft drive chain onto the crankshaft sprocket. The correct sprocket is the smaller of the two. If the starter clutch driven pinion was removed, it must be refitted at this stage. With the crankshaft held still, rotate the starter pinion in a clockwise direction so that the pinion boss enters the clutch body easily and does not displace the rollers.
4   Place a new oil seal on the left-hand crankshaft end with the spring side facing inwards. The crankshaft can now be lowered into place, making certain that the cam chain falls into the tunnel between the two crankcase mouths, and that the oil seal sits in its groove in the crankcase.

38.3a Lubricate and check positioning of main bearing shells

38.3b Rotate starter gear pinion in a clockwise direction to engage rollers

38.3c Install the front balance shaft and chain

38.3d Secure the spring with the spring clip

39 Engine reassembly: replacing the gear selector mechanism

1   Lubricate the gearchange drum bearings and slide the drum into position in the casing. Position the single forward gear selector fork so that the guide pin engages with the central track in the change drum. Slide the selector rod into position, to locate the fork. Fit the two rear selector forks in a similar manner.

40 Engine reassembly: replacing the gearbox components

1   If the gearshaft assemblies were dismantled for inspection or renovation they must now be assembled as complete units before fitting onto the casing. When reassembling either shaft, refer to the appropriate line drawing to ensure that all pinions, circlips and washers are refitted in **exactly** the correct order. Failure to do this will result in continuous gear selection problems. When fitting the bush upon which the mainshaft 6th gear pinion rotates, check that the hole in the bush aligns with

the hole drilled in the shaft.
2   Install the oil feed jet in the orifice of the mainshaft needle roller bearing housing in the crankcase lower half. Similarly, fit the layshaft needle roller bearing locating pin.
3   Install the completed mainshaft in the casing so that the dowel(s) locate with the bearings and the selector fork engages with the selector dog channel. Fit the completed layshaft in the same way. To ascertain whether the bearings have located correctly with their dowels, check that the two small scribed marks on the face of the outer races are in line with the casing mating surface.

41 Engine reassembly: replacing the cam chain guides and the mainbearing support casing

1   Slide the cam chain tensioner blade into position in the casing so the projections at the lower end locate in the recesses in the casing. Ensure that the bowed face is pointing towards the front of the engine. Position the balance chain guide and secure it by means of the single bolt. Check that the oil feed jet orifices are not blocked and then install the jet in the casing.

2   To ensure that engine balance is maintained, the balance weights must be timed in relation to the crankshaft. This is done by aligning given marks on the two weights and the crankshaft with various mating surfaces of the crankcase half. Rotate the crankshaft so that the 'B' mark on the extreme left-hand crankshaft web is aligned with the mating face of the crankcase half, towards the rear of the casing. Mesh the rear balance weight with the drive chain and rotate the assembly so that the 'TC' mark on the front balance weight is in line with the casing mating surface and the 'TH' mark on the rear balance weight is roughly in parallel with the mating surface. The centre run of the chain will now be meshed with the drive sprocket on the crankshaft. Final timing adjustment can take place only after the support casting has been fitted and the rear balance weight located on its shaft.

3   Install the main bearing shells in the support casting and check that the six hollow dowels are in place. Position the balance chain guide block in approximately the correct position in relation to the chain. Lubricate the main bearing journals and shells with clean engine oil and then lower the support casing into position.

4   Lift up the rear balance weight and slide the balance shaft into position so that the weight is temporarily secured. The balance weights should now be checked for timing by referring to the accompanying illustration. If the rear balancer is incorrectly positioned, withdraw the shaft and move the sprocket the required number of teeth on the chain. Refit the shaft and recheck the timing.

5   Install the six support casting main securing bolts and tighten them evenly, in a diagonal sequence, to a torque setting of 3·3 – 3·7 kgf m (24 – 27 lbf ft).

6   Check that the O-ring is fitted to the oil strainer union end and position the strainer on the support casting. Install the rearmost securing bolt, noting that the balance shaft may have to be rotated slightly – by inserting a screwdriver in the slot provided – so that the relieved portion of the shaft aligns with the bolt hole. Fit the two chain guide block securing bolts and final 6 mm support casing bolt. Tighten these bolts to a torque setting of 1·0 – 1·4 kgf m (7 – 10 lbf ft).

7   Before continuing, make a final check on the accuracy of the balance weight/crankshaft timing and check that all the shafts rotate freely in their bearing.

8   The balance chain may now be adjusted. With the adjuster nut slack, move the adjuster quadrant against the spring tension and allow it to flick back. The spring will automatically position the shaft to give the correct tension. Tighten the locknut.

39.1a Slide the change drum into position and ...

39.1b ... locate the front selector fork with the rod

39.1c Position the rear forks and locate with rod

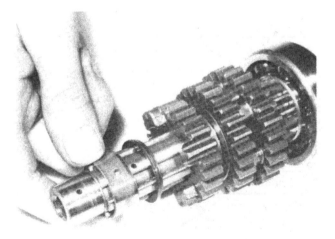

40.1 Ensure mainshaft 6th gear bush oil hole aligns with shaft hole

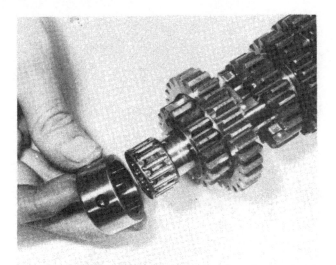

40.2a Mainshaft needle roller bearing and ...

40.2b ... layshaft needle roller bearing

40.2c Install oil feed jet for the mainshaft and ...

40.2d ... the layshaft bearing locating dowel

40.3a Fit layshaft oil seal plate and ...

40.3b ... the oil seal itself, as shown

40.3c The layshaft assembly fitted and ...

40.3d ... the layshaft and mainshaft assemblies fitted. Ensure bearings locate with pins

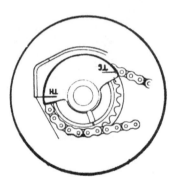

Rear balance weight

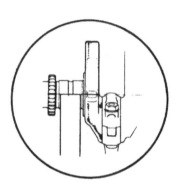

Crankshaft

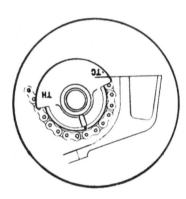

Front balance weight

Fig. 1.17 Balance weight timing

41.1 Install the chain guide block held by one bolt and install the oil feed jet (arrowed)

41.2a The front balance TC mark parallel with the casing

41.2b The TH mark on the rear balance weight roughly in parallel with the casing mating surface

41.2c Crankshaft timing mark below B on web

41.2d Mesh the rear balance weight with the chain in approximately the correct position

41.2e Place the chain guide block in approximately the correct position

41.3 Lower support casting into place. Check dowel positioning

41.5a Install support casting main securing bolts. Torque down evenly in a diagonal sequence

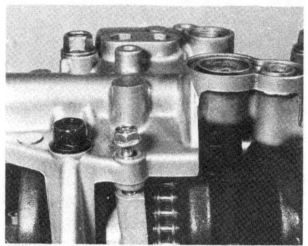

41.5b Secure chain guide block by single 6 mm bolt

42 Engine reassembly: replacing the kickstart shaft and starter idler pinion

1   The components listed in the heading must be refitted in the crankcase lower half before the crankcase halves can be rejoined.

2   Position the starter idler pinion in the casing recess so that the smaller diameter pinion is towards the centre of the casing. Check that the O-ring is in place on the pinion shaft and slide the shaft into position, so that it locates the pinion. The shaft should be inserted so that the cross-drilled hole at the inner end aligns with the bolt hole in the casing. Insert and tighten the single retaining bolt.

3   If previously removed, the oil pressure relief valve should now be refitted to the lower crankcase on later models.

N models only

4   Place the kickstart pinion in the casing, with the plain face towards the inside. Slip the heavy thrust washer between the pinion and the spindle support lug and then insert the spindle to locate the two components. Secure the spindle by means of the E-clip. Reassembly of the remaining kickstart components may be accomplished now or at a later stage, as described in Section 45.

42.2a Do not omit O-ring when fitting starter pinion and shaft

42.2b Securing bolt passes through hole in shaft

42.3a With kickstart pinion and shaft inserted ...

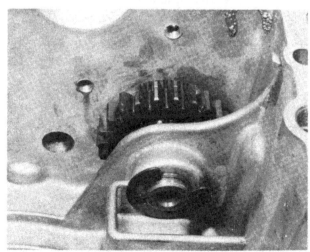

42.3b ... it is secured with the E-clip

43 Engine reassembly: joining the crankcase halves

1   Fit a new figure of eight O-ring to the recess in the support casting, and then check that the dowels are fitted to either the upper or lower crankcase half.
2   Make sure that the mating surfaces of the two crankcase halves are perfectly clean. Apply gasket compound to the face of the lower crankcase half. The lower casing can now be fitted over the upper crankcase and all the components. It is prudent to lubricate the various bearings with engine oil before fitting the lower casing. This will ensure that there is no oil starvation, however brief, when the engine is first run.
3   Insert all the retaining bolts into the crankcase and then tighten them in an even and diagonal sequence, to the following torque settings.

8 mm bolts   2·0 – 3·0 kgf m (15 – 22 lbf ft)
6 mm bolts   1·0 – 1·4 kgf m (7 – 10 lbf ft)

The tightening operation should be done in at least two stages.

44 Engine reassembly: replacing the oil filter and starter motor

1   Insert the oil filter centre bolt through the chamber cover and fit the compression spring and backing washer. Push a new oil filter element onto the centre bolt and install the sub-assembly on the crankcase. Note that the chamber cover is provided with a location lug, to aid positioning. The forked lug must engage correctly with the projection on the casing before the bolt is tightened. If there is any doubt about the condition of the chamber sealing ring or the centre bolt O-ring, the seal in question should be renewed.
2   Lubricate the starter motor boss O-ring and slide the starter motor into position in the crankcase. If necessary, rotate the crankshaft slightly to enable the starter motor pinion to engage with the idler pinion. Fit and tighten the two bolts.

45 Engine reassembly: replacing the gear change mechanism and kickstart assembly

1   Position the gear change drum locating plate in the primary drive chamber and fit the countersunk retaining screws. Honda recommended that the screws be staked in position after tightening, to prevent the screws becoming loose in service. A locking fluid applied to the threads may be used as an alternative.
2   The remaining kickstart components should now be fitted. Fit the thrust washer first, followed by the ratchet piece, coil spring and the second thrust washer. Note that the ratchet piece must be fitted to the kickstart shaft in one position only; that is with the punch mark on each component in alignment. If this is not done incorrect kickstart operation will result. Slide the return spring onto the shaft and locate the inner turned end in the cross-drilled hole. Grasp the outer turned end with a stout pair of pliers and tension the spring in a clockwise direction until the hooked end can be secured on the anchor lug. With the spring in place, insert the spring guide, ensuring it is pushed fully home.
3   Fit the gearchange drum stopper arm into the casing and replace the change pins and pin securing plate so that the roller end engages with the change pin plate. Ensure that the shouldered end of the distance sleeve engages correctly with the stopper arm, and that when tightened fully the bolt does not prevent rotation of the arm. Check that the return spring is correctly located and tensioned.
4   Lightly grease the gearchange splined shaft and introduce it through the hole in the gearbox wall. Gently push the shaft home through the oil seal in the timing side wall. This must be done with care or the seal lip will be damaged by the splines on the shaft. While pushing the shaft home, hold back the change pawl, which is spring loaded, on the end of the main arm, so that it clears the change pin assembly. The centralising spring on the main arm must be positioned correctly, with one spring leg either side of the stop screw.
5   Screw the neutral indicator switch into the top of the gearbox.
6   Before continuing, check the selection of each gear in turn.

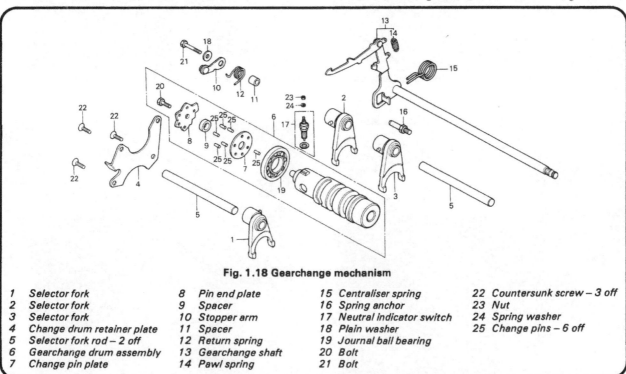

Fig. 1.18 Gearchange mechanism

1	Selector fork	8	Pin end plate	15	Centraliser spring
2	Selector fork	9	Spacer	16	Spring anchor
3	Selector fork	10	Stopper arm	17	Neutral indicator switch
4	Change drum retainer plate	11	Spacer	18	Plain washer
5	Selector fork rod – 2 off	12	Return spring	19	Journal ball bearing
6	Gearchange drum assembly	13	Gearchange shaft	20	Bolt
7	Change pin plate	14	Pawl spring	21	Bolt

22	Countersunk screw – 3 off
23	Nut
24	Spring washer
25	Change pins – 6 off

43.1a Fit a new double O-ring into the recesses

44.1 Check chamber cover O-ring before fitting oil filter

44.2 Lubricate the starter motor O-ring before replacing the motor

45.1 Position the gear change drum locating plate

45.5a Install the change drum stopper arm

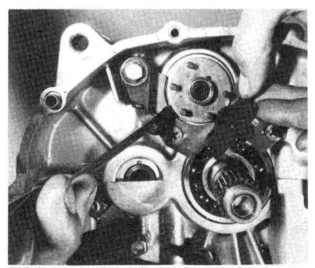
45.5b Replace the change pins and plate

45.6a Fit centraliser spring only as shown

45.6b Fit pawl arm, depress to clear change pin end plate. Note spring position

46 Engine reassembly: replacing the primary drive pinion and oil pump

1 On early models (with the oil pressure relief valve positioned beneath the oil pump) note that the valve can now be refitted. If dismantled reassemble the valve taking care not to overtighten it on refitting.

2 Check that the two hollow dowels are fitted to the rear of the oil pump. Place a new gasket on the rear of the pump and position the complete unit in the casing. Fit, and tighten evenly, the retaining screws, whilst simultaneously rotating the oil pump spindle to check that the pump casing does not distort and cause binding. If this does occur, remove the screws and check that the mating surfaces are absolutely clean.

3 Fit the internally splined washer over the crankshaft so that the chamfered inside diameter face is towards the engine. Slide the primary drive pinion onto the splined shaft.

4 The various components which comprise the oil pump drive should now be assembled and fitted. Fit the sprocket spacer onto the crankshaft with the large face against the primary drive pinion. Mesh the two pump sprockets with the chain and fit them onto their respective shafts so that the oil pump drive pin engages the rear of the driven sprocket and the drive sprocket engages with the two flats milled on the extreme end of the crankshaft.

5 Insert and tighten the centre bolt and secure the driven pinion by means of the circlip. To prevent crankshaft rotation when tightening the centre bolt, pass a close fitting bar through one small end eye in the same manner as that used for dismantling purposes.

47 Engine reassembly: Replacing the clutch assembly and primary drive cover

1 Place the heavy thrust washer over the clutch shaft and fit the clutch spacer bush. Lubricate the bush and fit the clutch outer drum so that the primary drive pinions engage correctly.

2 Place the clutch centre on the workbench so that the flanged face is downwards. Fit the cone spring seat, the cone spring and the backing plate, ensuring that the spring is fitted with the outer diameter away from the clutch centre. Secure these components by means of the spiral circlip. Fit the clutch plates to the clutch centre one at a time, starting with a friction plate and then a plain plate and a friction plate alternately. Line up the tangs of the friction plates so that they are level with

each other. Fit the pressure plate casting to the assembly and lift up the completed unit and insert it in the clutch outer drum. Some adjustment to the position of the clutch plate tangs may be necessary to enable the unit to be fitted.

3 Replace the clutch peg nut and the special domed washer. This has one face marked **outside** and should be fitted accordingly. Tighten the nut with the peg spanner used during dismantling, and prevent clutch rotation by adopting the same method.

4 Replace the clutch springs, thrust plate and bolts. The bolts should be tightened evenly, in a diagonal sequence, until tight.

5 Refit the oil pressure warning switch in the top of the casing. Note that the switch has a tapered thread, which aids oil tightness, and therefore will not screw in fully. Reconnect the switch lead held at the terminal by a single screw.

6 Place a new gasket on the outer face of the crankcase; the dowels will hold it in position. Position the cover and knock it gently onto the dowels. Replace the retaining bolts and tighten them down evenly in a criss-cross sequence. Take care when fitting the outer cover that the kickstart shaft splines do not damage the oil seal lip.

48 Engine reassembly: replacing the alternator and final drive sprocket

1 Position the alternator stator against the crankcase left-hand wall so that the three retaining screws can be inserted. Fit the stator so that the wiring leads are at the top. Press the ignition pick-up into place so that it locates correctly on the dowel, and insert the screws. Tighten the two sets of screws evenly, to avoid distortion of either components. Push the two wiring grommets into the cut-outs in the casing wall and secure the lead by means of the small clip held by a single screw.

2 Refit the Woodruff key into the recess in the tapered portion of the crankshaft, ensuring that it is properly located. Position the alternator rotor so that the keyway is in line with the key and push it fully home on the shaft. Fit and tighten the rotor centre bolt, preventing rotation of the crankshaft by using the close fitting bar passed through the small end eye of one connecting rod.

3 Place the final drive sprocket on the splined gearbox ouput shaft and then fit the securing plate. Rotate the plate so that the two bolt holes align with those in the sprocket and then fit and tighten the two bolts.

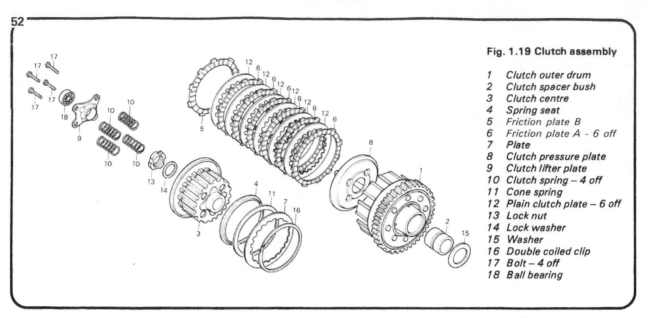

Fig. 1.19 Clutch assembly

1 Clutch outer drum
2 Clutch spacer bush
3 Clutch centre
4 Spring seat
5 Friction plate B
6 Friction plate A - 6 off
7 Plate
8 Clutch pressure plate
9 Clutch lifter plate
10 Clutch spring – 4 off
11 Cone spring
12 Plain clutch plate – 6 off
13 Lock nut
14 Lock washer
15 Washer
16 Double coiled clip
17 Bolt – 4 off
18 Ball bearing

46.2 Install the oil pump together with a new gasket and ...

46.3 ... fit thrust washer, pinion and sprocket spacer

46.4a Fit the two sprockets and chain simultaneously

46.4b Fit and tighten centre bolt; secure lower sprocket with circlip

47.1a Install the heavy thrust washer on the clutch shaft

47.1b Fit the clutch outer drum and spacer

47.2a Fit the clutch plates alternately, one at a time

47.2b Install the pressure plate and line up the plate tangs

47.3a Secure clutch temporarily by fitting spring, bolts and washers

47.3b Note marks on centre nut washer; fit accordingly

47.4 Replace the thrust plate, tightening the bolts evenly

47.5 Do not overtighten the oil pressure switch

47.6a Fit the primary drive cover and a new gasket ...

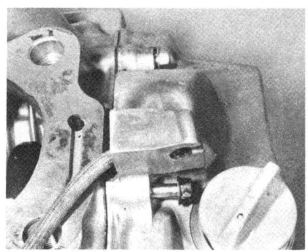

47.6b ... ensuring switch lead is positioned correctly

49 Engine reassembly: replacing the pistons, cylinder block and cylinder head

1   Before replacing the pistons, place a clean rag in each crankcase mouth to prevent any displaced component from falling into the crankcase. It is only too easy to drop a circlip while it is being inserted into the piston boss, which would necessitate a further strip down for its retrieval.

2   Before refitting the pistons the piston rings must be replaced. On 400cc models, the three-piece oil controlled ring should be fitted, one piece at a time, starting with a side rail, then the centre expander and then the final side rail. These components may be fitted either way up. The remaining rings, including the one-piece oil control ring fitted to 250cc models, should be fitted with the identification markings uppermost. If any difficulty is encountered in identifying the two top rings, it should be noted that the 2nd ring has a slightly tapered section and the top ring a parallel section, with slightly chamfered edges on the outer face.

3   Fit the pistons on to their original connecting rods, with the mark embossed on each piston crown facing forwards. If the gudgeon pins are a tight fit in the piston bosses, warm each piston first to expand the metal. Do not forget to lubricate the gudgeon pin, small end eye and the piston bosses before reassembly.

4   Use new circlips. **NEVER** re-use old circlips. Check that each circlip has located correctly in its groove. A displaced circlip will cause severe engine damage.

5   Fit a new cylinder base gasket over the holding down studs. Gasket compound must not be used. Insert the two locating dowels in the front outer cylinder head bolt holes and check that the two oil feed jets in the mating surface are in position and not clogged. Fit new O-rings to each cylinder sleeve where they project from the cylinder barrel.

6   Position the three oil control ring components on each piston (400 models) so that the end gaps are at least 20 mm (0.78 in) away from one another.

7   Position the piston rings so that their end gaps are out of line with each other and fit a piston ring clamp to each piston. It is highly recommended that ring clamps be used for cylinder block replacement. This operation can be done by hand but is a tedious task and it is all too easy to fracture a ring. Position the piston at TDC.

8 The cylinder block replacement operation requires two persons; one to hold the pistons and help feed the pistons into the bores, and the other to support the cylinder block as it is lowered into position. Lubricate the cylinder bores and the piston rings thoroughly. Position the cylinder block above the pistons and, with a length of suitably bent wire, hook the camshaft chain up through the chain tunnel between the cylinder bores. Run a screwdriver or rod through the chain across the top of the barrel to prevent the camshaft chain falling free.

9 Carefully slide the cylinder block down until the pistons enter the cylinder bores. Keeping the pistons square to the bores ease the block down until the piston clamps are displaced. Remove the two ring clamps and the rag padding from the crankcase mouths and push the cylinder block down onto the base gasket.

10 Fit a new O-ring to the stud which projects from the cam chain tensioner body. Install the body in the rear of the cylinder block and fit the washer and nut. Do not tighten the nut at this stage. Reconnect the tensioner blade with the moving tensioner arm by inserting the clevis pin and fitting the R-spring clip. Take care not to drop any components down the cam chain tunnel. Pull the moving arm upwards so that it is extended fully and lock it in this position by tightening the adjuster nut. With the arm so placed, the tensioner blade has the flattest profile; this facilitates cam chain and camshaft reassembly.

11 Install the cam chain guide blade in the front of the cylinder block, ensuring that the lower end locates correctly in the casing.

12 Fit the four hollow dowels to the outer bolt holes in the cylinder block top mating surface, and place a new 'O' ring around each rear outer dowel. Place a new cylinder head gasket in position over the dowels.

13 Position the cylinder head above the engine and using the hooked wire, pull the camshaft drive chain up through the chain tunnel in the cylinder head and fit the head onto the dowels. Place a screwdriver through the chain again, across the cylinder head.

14 Insert and tighten the chain tensioner upper bolt; do not omit the O-ring.

49.3a Lubricate small-end bearing before installing piston and gudgeon pin

49.3b Fit piston with IN mark towards rear. EX mark (arrowed) forwards

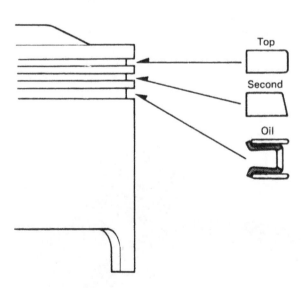

Fig. 1.20 Piston ring profiles

Top

Second

Oil

49.5 Insert the two locating dowels and position the two oil feed jets

49.8 Feed the camchain through the chain tunnel

49.9 When the rings are fully home remove the rag

49.10a Reconnect the camchain tensioner and blade

49.10b Pull the tensioner upwards, locking it by means of the nut

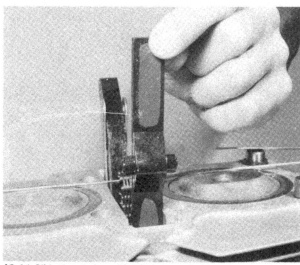
49.11 Slide the chain guide blade into place

49.12a Ensure the four dowels are refitted

49.12b Fit a new cylinder head gasket before ...

49.13 ... lowering the head into position

49.14 Fit the camchain tensioner upper bolt. Note the O-ring

50 Engine reassembly: replacing the camshaft, timing the valves and fitting the rocker gear

1   Secure the cam chain, using a length of thin wire so that during camshaft replacement the chain will not fall into the tunnel. Position the camshaft sprocket to the left of the chain and insert the camshaft from the right-hand side. The timing marks scribed on the sprocket face must face towards the left-hand side. To allow the camshaft centre boss to enter the sprocket centre, the sprocket must be positioned with the off-set cutaways approximately vertical. With the sprocket positioned against the flange, rotate the camshaft until the slot machined in the left-hand end is in the 12 o'clock position and rotate the sprocket so that the opposed scribe marks are parallel with the cylinder head. Apply a spanner to the alternator rotor centre bolt and rotate the rotor in a forward direction until the 'T' mark on the rotor is in exact alignment with the raised index mark on the casing. Now lift the cam chain onto the sprocket and align the bolt holes in the sprocket with those in the camshaft flange.

2   Fit and partially screw in the first sprocket bolt. If the timing is correct, the 'T' mark on the alternator rotor must be in line with the index mark and the two scribed lines on the sprocket must be parallel with the cylinder head cover mating surface. In addition, the milled slot in the camshaft left-hand end must be in the 12 o'clock position. If the marks do not align exactly, repeat the timing sequence and re-check. Absolute accuracy is essential.

3   Apply non-permanent locking fluid to the sprocket bolts and tighten them to a torque setting of 1.8 - 2.2 kgf m (13 - 16 lbf ft). The crankshaft must be rotated to insert the second bolt, and the bolts must be tightened evenly.

4   Lubricate thoroughly the camshaft journals and lobes and fill the troughs below each lobe with engine oil. Similarly, lubricate the valve stems and springs. Place the assembled rocker carriers into position on the cylinder head, after checking that the hollow dowels are fitted in the four outer bolt holes. Insert the long bolts that pass through the carriers, cylinder head and cylinder block and screw into the crankcase. Note that the four centre bolts are fitted with a copper washer under their heads. The rocker pin securing plates fitted to the outer two bolts on each carrier are not interchangeable.

5   Before tightening the bolts, note that with the engine positioned with the pistons at TDC, one cam lobe will be sufficiently raised to actuate the adjacent rocker arm. To prevent distortion, slacken off the rocker adjuster on the relevant rocker arm, after loosening the locknut. Tighten the bolts evenly, in a number of steps, in a reversal of the sequence given in the Fig. 1.3. The correct torque setting is 3.0 - 3.3 kgf m (22 - 24 lbf ft).

6   Before the cylinder head cover is replaced, the clearance between each rocker arm and valve stem must be checked and, if required, adjusted. With the pistons at TDC (when the rotor 'T' mark is aligned with the index mark) the valves for one or other cylinder will be closed, with the cam lobes pointing downwards. Check the clearances on these three valves, using a feeler gauge of the correct size. The recommended clearances are as follows:

	CB250 N	CB400 N
Inlet	0.12 mm (0.005 in)	0.10 mm (0.004 in)
Exhaust	0.16 mm (0.006 in)	0.14 mm (0.005 in)

If required, loosen the locknut and turn the adjuster until the feeler gauge is a light sliding fit. Tighten the locknut, without allowing the adjuster to turn, and re-check the gap. Rotate the crankshaft through 180° until the pistons are again at TDC. Check and adjust the valve clearances on the second cylinder.

7   Make sure that the cylinder head cover sealing ring is in good condition and correctly positioned and refit the cylinder head cover. Secure the cover by means of the special bolts and rubber washers. If the condition of the washers is suspect, they should be renewed.

8   Rotate the crankshaft anti-clockwise until the 'T' mark on

50.1 Feed the camshaft through the sprocket and chain as shown

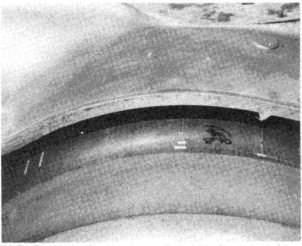

50.2a Cam timing is correct when T mark on alternator and ...

50.2b ... parallel lines on sprocket are as shown and the camshaft slot is in the 12 o'clock position

50.4 Fit the assembled rocker carriers. Note that the four centre bolts have copper washers

the rotor is in line with the index mark. Slacken the cam chain tensioner locknut and then tighten it. The tensioner will now be set correctly for proper chain adjustment.

51 Replacing the engine/gearbox unit in the frame

1   As is the case with removal, engine replacement requires considerable care and patience. The initial stage of replacement requires two people; one to guide the engine, and the other to operate the jack. Lift the engine on to the jack so that it is as well balanced as possible. Raise the engine slowly so that the rear of the casing enters the mounting plates squarely. Align the two rear mounting bolt holes, tilting the engine as necessary, and insert but do not tighten the bolts. Refit the engine front mounting bracket, noting that the starter motor cable runs down inside the bracket. Replacement of the head steady bracket should be carried out only after the carburettors have been refitted. Fit the bolts from the left-hand side of the machine in the interest of uniformity and then tighten the nuts fully. Note that the engine rear upper mounting bolt nut secures the earth cable from the battery. This must not be omitted.

2   Loop the final drive chain over the engine sprocket and mesh the two ends onto the rear wheel sprocket. Replace the

50.6 The feeler gauge should be a light sliding fit

master link making certain that the spring link is replaced the correct way round. That is with the closed end facing the direction of travel. Replace the lower chain protector plate. Fit the final washer onto the gearchange shaft. Slide the final drive sprocket cover into position and secure it by means of the bolts.

3 Re-install the rear brake pedal by reversing the dismantling procedure. Grease the pivot bolt and bush before inserting them. After reconnecting the brake operating rod to the arm on the brake back plate, adjust the brake by means of the adjuster nut, so that there is about 20 - 30 mm ($\frac{3}{4}$ - $1\frac{1}{4}$in) movement at the brake pedal foot pad before the brake begins to bite.

4 Check that the air hose screw clips are in position on the two hoses. Place the carburettors as a unit to the right-hand side of the machine so that their control cables may be reconnected. Connect the closing cable first and then hold the throttle pulley in the open position and refit the opening cable to the rear. Adjust the cables by means of the lower adjusters so that there is a small amount of free play at the throttle twistgrip. Reconnect the choke cable inner to the choke arm and secure the outer cable at the clamp so that there is a small amount of free play. Push the carburettors into position so that the carburettor mouths are fully home in the air hoses. The inlet stubs should now be fitted one at a time, using the same system adopted during dismantling. Ensure that the O-ring in each inlet stub flange is in good condition and stands proud of the flange face when in the groove. If the O-ring is below the surface of the flange, it will serve no useful purpose and **MUST** be renewed.

5 With the carburettors installed, the head steady bracket may be replaced. Note that the rear bolt which holds the left-hand bracket plate to the frame also secures a wiring clamp. Reconnect the breather tube to the air filter box and cylinder head cover unions, securing the hose at each end by means of the spring clips.

6 Reconnect the alternator leads at the block connectors and snap connectors adjacent to the regulator/rectifier unit and reconnect the oil pressure and neutral indicator switch leads. Secure the starter motor cable at the motor terminal and pull down the rubber boot.

7 Reconnect the clutch cable to the operating arm, after passing it through the anchor bracket on the casing. Adjust the cable so that there is approximately 10 - 20 mm ($\frac{3}{8}$ - $\frac{3}{4}$ in) movement measured at the handlebar lever ball end before the clutch begins to lift. Do not forget to tighten the locknuts, after adjustment.

8 Insert the tachometer cable lower end into the take-off housing in the primary drive cover and fit and tighten the securing bolt. If difficulty is encountered engaging the drive dog with the cable, rotate the engine, using the kickstart.

9 Position the exhaust balance box below the crankcase so that the bolts may be inserted and partially tightened. Fit a new exhaust gasket ring into each of the two exhaust ports and replace the exhaust pipes individually. Replace the silencers. Tighten the exhaust flange nuts first, followed by the front clamps, balance box bolts etc. This will prevent any part of the system being stressed in position. Fit the rider's footrests.

10 Refit the petrol tank and connect the fuel lines. Do not omit the small spring clips.

11 Reconnect the plug caps to their respective sparking plugs and reconnect the battery terminals. Give a final visual check to all electrical connections and replace the two side covers. Both are a push fit.

12 Pour the recommended quantity and grade of oil into the engine through the filler orifice to the right of the cylinders. Kick the engine over smartly with the ignition off to help prime the oilways.

51.2 Refit the engine left-hand cover using a new gasket

51.4a Reconnect the throttle cables with the opening cable at the rear and ...

51.4b ... manoeuvre the carburettors into position

51.7a Reconnect the clutch cable with the operating arm

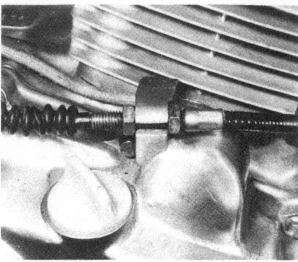

51.7b Adjust the clutch cable play at the adjuster

51.8 Re-insert the tachometer cable and retaining bolt

51.9a Support the balance box while inserting the securing bolts

51.9b Fit a new gasket ring in each exhaust port

51.12 Refill the engine with the correct quantity of oil

52 Starting and running the rebuilt engine

1 Open the petrol tap, close the carburettor chokes and start the engine, using either the kickstart or the electric starter. Raise the chokes as soon as the engine will run evenly and keep it running at a low speed for a few minutes to allow oil pressure to build up and the oil to circulate. If the red oil pressure indicator lamp is not extinguished, stop the engine immediately and investigate the lack of oil pressure.

2 The engine may tend to smoke through the exhausts initially, due to the amount of oil used when assembling the various components. The excess of oil should gradually burn away as the engine settles down.

3 Check the exterior of the machine for oil leaks or blowing gaskets. Make sure that each gear engages correctly and that all the controls function effectively, particularly the brakes. This is an essential last check before taking the machine on the road.

53 Taking the rebuilt machine on the road

1 Any rebuilt machine will need time to settle down, even if parts have been replaced in their original order. For this reason it is highly advisable to treat the machine gently for the first few miles to ensure oil has circulated throughout the lubrication system and that any new parts fitted have begun to bed down.

2 Even greater care is necessary if the engine has been rebored or if a new crankshaft has been fitted. In the case of a rebore, the engine will have to be run-in again, as if the machine were new., This means greater use of the gearbox and a restraining hand on the throttle until at least 500 miles have been covered. There is no point in keeping to any set speed limit; the main requirement is to keep a light loading on the engine and to gradually work up performance until the 500 mile mark is reached. These recommendations can be lessened to an extent when only a new crankshaft is fitted. Experience is the best guide since it is easy to till when an engine is running freely.

3 If at any time a lubrication failure is suspected, stop the engine immediately, and investigate the cause. If an engine is run without oil, even for a short period, irreparable engine damage is inevitable.

4 When the engine has cooled down completely after the initial run, recheck the various settings, especially the valve clearances. During the run most of the engine components will have settled into their normal working locations.

54 Fault diagnosis: engine

Symptom	Cause	Remedy
Engine will not start	Defective sparking plugs	Remove the plugs and lay them on cylinder head. Check whether spark occurs when ignition is switched on and engine rotated.
	Faulty CDI ignition	Have ignition components checked. Renew if faulty.
Engine runs unevenly	Ignition and/or fuel system fault	Check each system independently, as though engine will not start.
	Blowing cylinder head gasket	Leak should be evident from oil leakage where gas escapes.
	Incorrect ignition timing	Check accuracy and if necessary reset.
Lack of power	Fault in fuel system or incorrect ignition timing	See above.
Heavy oil consumption	Cylinder block in need of rebore	Check for bore wear, rebore and fit oversize pistons if required.
	Damaged oil seals	Check engine for oil leaks.
Excessive mechanical noise	Worn cylinder bores (piston slap)	Rebore and fit oversize pistons.
	Worn camshaft drive chain (rattle)	Adjust tensioner or replace chain.
	Worn big-end bearings (knock)	Fit replacement crankshaft assembly.
	Worn main bearings (rumble)	Fit new journal bearings and seals. Renew crankshaft assembly if centre bearings are worn.
Engine overheats and fades	Lubrication failure	Stop engine and check whether internal parts are receiving oil. Check oil level in crankcase.

Clutch and gearbox 'Fault diagnosis' on page 62

55 Fault diagnosis: clutch

Symptom	Cause	Remedy
Engine speed increases as shown by tachometer but machine does not respond	Clutch slip	Check clutch adjustment for free play at handlebar lever. Check thickness of inserted plates.
Difficulty in engaging gears. Gear changes jerky and machine creeps forward when clutch is withdrawn. Difficulty in selecting neutral.	Clutch drag	Check clutch adjustment for too much free play. Check clutch drums for indentations in slots and clutch plates for burrs on tongues. Dress with file if damage not too great.
Clutch operation stiff	Damaged, trapped or frayed control cable	Check cable and renew if necessary. Make sure cable is lubricated and has no sharp bends.

56 Fault diagnosis: gearbox

Symptom	Cause	Remedy
Difficulty in engaging gears	Selector forks bent Gear clusters not assembled correctly	Renew. Check gear cluster arrangement and position of thrust washers.
Machine jumps out of gear	Worn dogs on ends of gear pinions Stopper arms not seating correctly	Renew worn pinions. Remove right-hand crankcase cover and check stopper arm action.
Gearchange lever does not return to original position	Broken return spring	Renew spring.
Kickstart does not return when engine is turned over or started	Broken or poorly tensioned return spring	Renew spring or re-tension.
Kickstart slips	Ratchet assembly worn	Part crankcase and renew all worn parts.

Chapter 2 Fuel system and lubrication

Contents

Specifications

Fuel tank capacity

	Overall	Reserve
	14 lit	3.5 lit
	(3.7/3.1 US/Imp gall)	(0.9/0.8 US/Imp gall)

Carburettors

	250N	400N
Make	Keihin	Keihin
Type	VB30A (28 mm)	VB31A (32 mm)
Air screw opening	2 turns	$1\frac{1}{2}$ turns
Float height	15.5 mm (0.61 in)	15.5 mm (0.61 in)
Vacuum (at idle)	170–210 mm Hg	200–240 mm Hg
Idle speed	1300 ± 100 rpm	1200 ± 100 rpm

Engine/transmission oil

All models

Capacity:
- At oil change .... 2.2 litre (3.9 pint)
- At oil and filter change .... 2.3 litre (4.0 pint)
- Dry .... 3.0 litre (5.3 pint)
- Oil grade .... SAE 10W/40, 15W/40 or 20W/50

Oil pump

- Type .... Trochoid
- Inner rotor/outer rotor clearance .... 0.10 mm (0.004 in)
- Outer rotor/body clearance .... 0.35 mm (0.014 in)
- Rotor end float .... 0.10 mm (0.004 in)
- Oil pressure .... 56.9–75.4 psi (4.0–5.3 kg/cm²)

1 General description

The fuel system comprises a petrol tank from which petrol is fed by gravity to the float chambers of the two Keihin carburettors. A single fuel tap with detachable gauze filter is located beneath the tank, on the left-hand side. It contains provision for a reserve petrol supply when the main supply is exhausted.

For cold starting, a car-type choke control is provided, interconnected by a cable to the linked choke butterflies in the carburettors so that the mixture of both carburettors can be enriched temporarily. Throttle control is effected by the means of a push-pull cable arrangement which operates a 'butterfly' valve in each carburettor.

Lubrication is by the wet sump principle in which oil is delivered under pressure, from the sump, through a mechanical pump to the working parts of the engine. The pump is of the trochoid type and is driven from the crankshaft by a small single row chain. Oil is supplied under pressure via a release valve and full flow oil filter fitted with a paper element. The engine oil is also shared by the primary drive and the gearbox.

2   Petrol tank: removal and replacement

1   The fuel tank is retained at the forward end by two cups, one of which is placed each side of the underside of the tank, which locate with two rubber buffers on the frame top tube. The rear of the tank rests on a small rubber saddle placed across the frame top tube, and is retained by a single bolt passing through a lug projecting from the tank.

2   To remove the tank, pull off the fuel line at the petrol tap, after turning the top lever to the 'off' position. Release the pressure on the fuel line by pinching together the ears of the securing spring clip. Detach the dualseat from the machine by depressing simultaneously the two spring loaded catches located below the rear of the seat. After removing the single securing bolt, the tank can be pulled backwards, until the cups clear the rubber buffers, and lifted upwards away from the machine.

3   When replacing the tank, reverse the above procedure. Make sure the tank seats correctly and does not trap any control cables or wires. If difficulty is encountered when trying to slide the tank cups onto the rubbers, a small amount of petrol may be applied to the rubbers as a lubricant.

3   Petrol tap: removal and replacement

1   The petrol tap is retained on the underside of the tank by a gland nut which is concentric with the tap body. Removal of the tap, either for filter cleaning or renewal, must be preceded by draining the tank. This can be done most easily by fitting a suitable length of tubing, the lower end of which is placed in a container.

2   Unscrew the gland nut fully and withdraw the top from the tank. The filter column may be slid off the stand pipe, for clean-

ing. This should be done using a soft brush and clean petrol. If the filter screen has become perforated, it should be renewed.

3   The tap is a sealed unit; therefore, if leakage occurs at the lever the complete assembly must be renewed.

4   When fitting the tap to the petrol tank, apply a small amount of sealing compound to the gland nut threads.

4   Petrol feed pipe: examination

1   A synthetic rubber feed pipe is used to convey the flow of fuel from the petrol tap to the float chamber union of the left-hand carburettor. The pipe is retained at each end by a wire clip, which must hold the pipe firmly in position. Check periodically to ensure the pipe has not begun to split or crack and that the wire clips have not worn through the surface.

2   Do NOT replace a broken pipe with one of natural rubber, even temporarily. Petrol causes natural rubber to swell very rapidly and disintegrate, with the result that minute particles of rubber would easily pass into the carburettors and cause blockages of internal passageways. Plastic pipe of the correct bore size can be used as a temporary substitute but it should be replaced with the correct type of tubing as soon as possible since it will not have the same degree of flexibility.

5   Carburettors: removal

1   Before attempting to detach the carburettors, the petrol tank should be removed to improve access as described in Section 2 of this Chapter.

2   Because of the limited space between the rear of the cylinder head and the air filter box, carburettor removal requires some dexterity and care. In addition, the correct removal

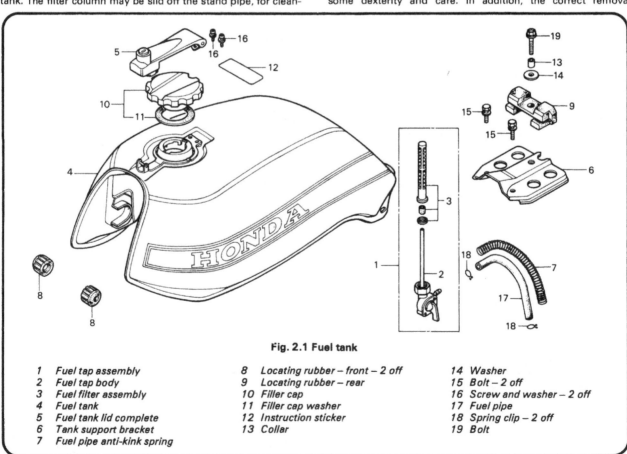

Fig. 2.1 Fuel tank

1	Fuel tap assembly	8 Locating rubber – front – 2 off	14 Washer
2	Fuel tap body	9 Locating rubber – rear	15 Bolt – 2 off
3	Fuel filter assembly	10 Filler cap	16 Screw and washer – 2 off
4	Fuel tank	11 Filler cap washer	17 Fuel pipe
5	Fuel tank lid complete	12 Instruction sticker	18 Spring clip – 2 off
6	Tank support bracket	13 Collar	19 Bolt
7	Fuel pipe anti-kink spring		

sequence must be followed. Removal of the head steady bracket, the plates of which pass between the two carburettors, is not strictly necessary, although it will improve access.

3 Slacken fully the screw clips holding the carburettors to the air filter box stubs and the inlet stubs. Slide the air hose clips to the rear so that they clear the carburettor mouths. Remove the two bolts securing each inlet stub to the cylinder head.

4 Grasp firmly the right-hand carburettor and pull it rearwards, so that the air hose concertinas. This will give sufficient clearance to enable the inlet hose to be pulled off the carburettor. Repeat this operation on the left-hand instrument. Pull the carburettors forward, as a unit, out of the air hoses and then out towards the right-hand side of the machine.

5 Before the carburettors are completely free for removal, the control cables must be detached. Slacken the choke cable clamp on the carburettor so that the inner cable can be disconnected from the operating arm and the complete cable pulled from position. Slacken the adjuster locknut on the lower cable adjuster of each throttle cable and screw in the adjusters to increase play in the cable. Open the throttles by rotating the pulley and disconnect the rear cable (opening cable) from the pulley. Close the throttle and disconnect the front cable (closing cable).

2.2 Single bolt retains fuel tank at rear

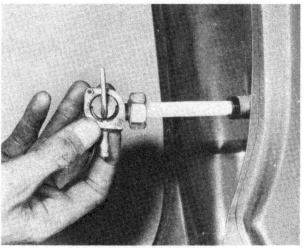

3.2 Withdrawing the petrol tap and filter

5.4a Carburettors can be removed as a unit and ...

5.4b ... the throttle cables detached after removal

5.4c The carburettors removed as a unit

6 Carburettors: dismantling, examination and reassembly

1   Before dismantling of the carburettors can be commenced, the two instruments must be separated from the mounting bar and the controls disconnected as follows:

2   Unhook the light spring that returns the choke link arm where the choke shafts interconnect at the left-hand instrument. Take special note of the position of the spring, to prevent confusion during reassembly. Loosen the two screws which hold each carburettor cap and remove the screw from each cap that secures the top bracket. The carburettors share a common mounting bar, held to each instrument by two screws. Remove the screws and detach the bracket. These screws may be very tight; take care not to damage the soft screw heads. The two instruments can now be pulled apart. As this is done, the choke link rod will disconnect from the choke arm on the left-hand instrument, and the throttle link rod arm will slide out of position. Note the coil spring placed concentrically between the ends of the throttle rods. This will fall free. The fuel transfer tube is a push fit in each carburettor, the seal being made by an O-ring at each end.

3   It is suggested that each carburettor is dismantled and reassembled separately, to avoid mixing up the components. The carburettors are handed and therefore components should not be interchanged.

4   Invert one carburettor and remove the three float chamber screws. Lift the float chamber from position and note the chamber sealing ring. This need not be disturbed unless it is damaged. The two floats, which are interconnected, can be lifted away after displacing the pivot pin. The float needle is attached to the float tang by a small clip. Detach the clip from the tang and store the needle in a safe place.

5   Pull the rubber blanking plug from the slow jet housing and unscrew the slow jet. When unscrewing any jet, a close fitting screwdriver must be used to prevent damage to the slot in the soft jet material. Hold the secondary main jet holder with a small spanner and unscrew the secondary jet. The holder may then be unscrewed to release the needle jet which is a push fit and projects into the carburettor venturi. Unscrew the main jet from the final housing and then unscrew the main nozzle from the same housing.

6   Unscrew the remaining screw which holds the carburettor cap (piston chamber) and pull the cap from position. Remove the helical spring and the nylon sealing ring. Pull the piston up and out of its slider. The piston needle can be removed by unscrewing the plug in the top of the piston. The needle will drop out. The main air jet and secondary air jet are hidden below a

plate, which is retained in the upper chamber by a single cross head screw. Remove the screw and plate. The two slow air jets are similarly positioned opposite the main air jets, but are not closed by a plate. None of these jets can be removed. They must be cleaned in place.

7   On 400 cc models an air cut-out valve is included in each carburettor to automatically richen the mixture when the throttle is closed suddenly after the engine has been running fast. This prevents backfiring in the exhaust system. The cut-out valve is enclosed by a cover held on the outside of the carburettor body by two screws. Unscrew the screws, holding the cover in place against the pressure of the diaphragm spring, and then lift the cover away. Remove the spring, and carefully lift out the diaphragm.

8   Check the condition of the floats. If they are damaged in any way, they should be renewed. The float needle and needle seating will wear after lengthy service and should be inspected carefully. Wear usually takes the form of a ridge or groove, which will cause the float needle to seat imperfectly. If damage to the seat has occured the carburettor body must be renewed because the seat is not supplied as a separate item.

9   After considerable service the piston needle and the needle jet in which it slides will wear, resulting in an increase in petrol consumption. Wear is caused by the passage of petrol and the two components rubbing together. It is advisable to renew the jet periodically in conjunction with the piston needle.

10   Inspect the cut-out valve diaphragm for signs of perishing or perforation. Damage will be easily seen.

11   Before the carburettors are reassembled, using the reversed dismantling procedure, each should be cleaned out thoroughly using compressed air. Avoid using a piece of rag since there is always risk of particles of lint obstructing the internal passageways or the jet orifices.

12   Never use a piece of wire or any pointed metal object to clear a blocked jet. It is only too easy to enlarge the jet under these circumstances and increase the rate of petrol consumption. If the compressed air is not available, a blast of air from a tyre pump will usually suffice.

13   Do not use excessive force when reassembling a carburettor because it is easy to shear a jet or some of the smaller screws. Furthermore, the carburettors are cast in a zinc-based alloy which itself does not have a high tensile strength. Take particular care when replacing the throttle valves to ensure the needles align with the jet seats.

14   Do NOT remove either the throttle stop screw or the pilot jet screw without first making note of their exact positions. Failure to observe this precaution will make it necessary to resynchronise both carburettors on reassembly.

6.2a Detach the mounting bar

6.2b Disconnect the light choke link arm return spring and ...

6.2c ... separate the two carburettors

6.4a The float chamber lifts off after the three screws are removed

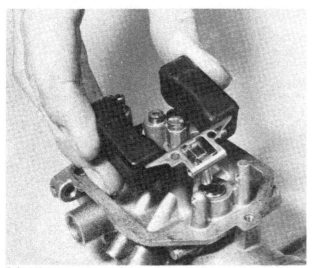

6.4b Lift the two floats clear as a unit

6.5a The slow jet is covered by a rubber plug

6.5b The main jet (arrowed) can be unscrewed

6.5c Secondary jet (arrowed) screws into housing which holds needle jet in place

6.6a Release cap to gain access to the piston

6.6b Withdraw the piston and needle

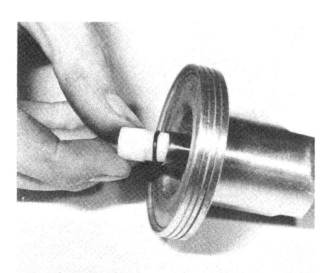

6.6c Pull out the nylon sealing plug ...

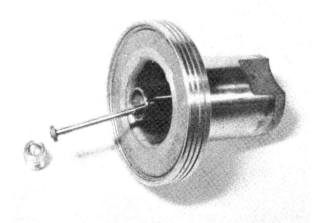

6.6d ... and unscrew the grub screw to allow needle removal from the piston

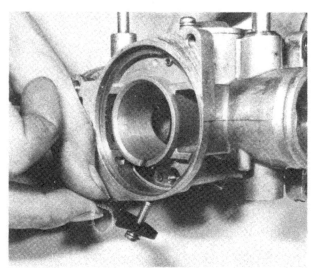

6.6e The air jets are covered by a plate retained by one screw

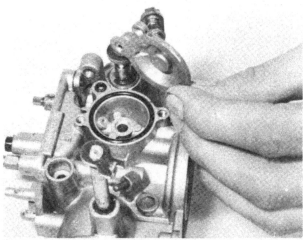

6.7 Cover retained by two screws which houses the cut-out valve on 400N model

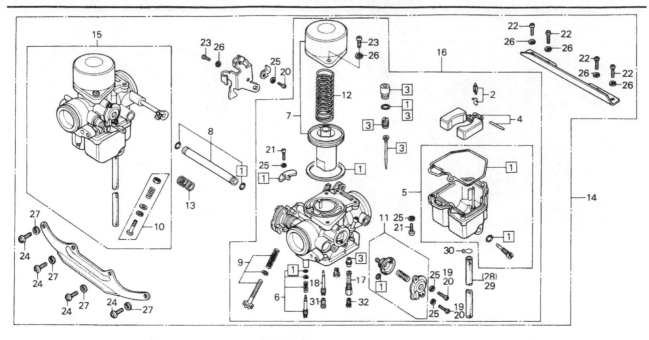

Fig. 2.2 Carburettors

1 Gasket set	11 Air cut-off valve assembly – 2 off	22 Screw – 4 off
2 Float needle/clip – 2 off	(not 250N)	23 Screw – 5 off
3 Jet needle assembly – 2 off	12 Spring – 2 off	24 Screw – 4 off
4 Float assembly – 2 off	13 Throttle link spring	25 Spring washer – 5 off
5 Float chamber assembly – 2 off	14 Carburettor assembly	26 Spring washer – 6 off
6 Pilot adjuster screw assembly	15 RH carburettor	27 Spring washer – 4 off
– 2 off	16 LH carburettor	28 Fuel drain pipe
7 Piston assembly – 2 off	17 Needle jet holder – 2 off	29 Fuel drain pipe
8 Fuel cross-feed piper	18 Main nozzle – 2 off	30 Spring clip – 4 off
9 Throttle stop screw assembly	19 Screw – 2 off	31 Main jet
10 Synchronization adjustment screw	20 Screw – (5 off – 400N)	32 Secondary jet
assembly	21 Screw – 3 off	

7 Carburettors: synchronisation

1 For the best possible performance it is imperative that the carburettors are working in perfect harmony with each other. At any given throttle opening if the carburettors are not synchronised, not only will one cylinder be doing less work but it will also in effect have to be carried by the other cylinder. This effect will reduce the performance considerably.

2 It is essential to use a vacuum gauge set consisting of two separate dial gauges, one of each being connected to each carburettor by means of a special adaptor tube. The adaptor pipe screws into the top of each inlet tract to the rear of the cylinder head, the orifice of which is normally blocked by a crosshead screw. Most owners are unlikely to possess the necessary vacuum gauge set, which is somewhat expensive and is normally held by Honda Service Agents, who will carry out the synchronising operation for a nominal sum.

3 If the vacuum gauge set is available, proceed as follows: Remove the dualseat and petrol tank so that access can be gained to the carburettors. Using a suitable length of feed pipe, reconnect the petrol tank with the carburettors, so that the petrol flow can be maintained. The petrol tank must be placed above the level of the carburettors. Connect the vacuum gauges to the engine.

4 Start the engine and allow it to run until normal working temperature has been reached. This should take 10–15 minutes. Set the throttle so that an engine speed of 1100 – 1300 rpm is maintained. If the readings on the vacuum gauges are not within 40 mm Hg of each other, some adjustment is required. Loosen the locknut securing the adjuster screw on the

throttle link shaft arm.

5 Turn the adjuster screw anti-clockwise to increase the reading on the right-hand carburettor, or clockwise to decrease the reading, until the vacuum levels for each carburettor are the same. Tighten the adjuster locknut and re-check.

6 After carrying out the synchronisation of carburettors the vacuum reading at tick-over should be 200 – 240 mm Hg (400 cc models) or 170 – 210 mm Hg (250 cc model).

8 Carburettors: adjustment for tickover

1 The carburettors should be adjusted for engine tick-over whenever carburettor synchronisation has been carried out or if rough idling is experienced. Before adjusting the carburettors a check should be made to ensure that the following settings are correct: contact breaker gap, ignition timing, valve clearance, sparking pluggaps, crankcase oil level. It is also important that the engine is at normal running temperature.

2 Start the engine and bring it up to normal running temperature. Unscrew the throttle stop screw, which is synchronised to both carburettors and is mounted on a bracket inboard of the left-hand carburettor, until the engine turns over at the slowest smooth speed obtainable. Turn each pilot screw an equal amount each until the engine reaches the highest rpm obtainable. Adjust the engine speed, using the throttle screw, until the engine is turning over at 1100 – 1300 rpm as shown by the tachometer.

3 Alter the position of the pilot screws an equal amount to check whether the engine speed rises. If this is the case repeat the procedure given in the previous paragraph.

8.2 Pilot screw adjuster

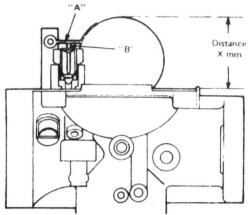

Fig. 2.3 Checking float level

A   Float tongue
B   Float valve
X   =   15.5 mm (0.61 in)

9 Carburettor settings

1   Some of the carburettor settings, such as the sizes of the needle jets, main jets and needle positions, etc, are predetermined by the manufacturer. Under normal circumstances it is unlikely that these settings will require modification, even though there is provision made. If a change appears necessary, it can often be attributed to a developing engine fault.
2   Always err slightly on the side of a rich mixture, since a weak mixture will cause the engine to overheat. Reference to Chapter 3 will show how the condition of the sparking plugs can be interpreted with some experience as a reliable guide to carburettor mixture strength. Flat spots in the carburation can usually be traced to a defective timing advancer. If the advancer action is suspect, it can be detected by checking the ignition timing with a stroboscope.

10 Carburettors: adjusting float level

1   If problems are encountered with fuel overflowing from the float chambers, which cannot be traced to the float/needle assembly or if consistent fuel starvation is encountered, the fault will probably lie in maladjustment of the float level. It will be necessary to remove the float chamber bowl from each carburettor to check the float level.
2   If the float level is correct the distance between the uppermost edge of the floats and the flange of the mixing chamber body will be 15.5 mm (0.61 in).
3   Adjustments are made by bending the float assembly tang (tongue), which engages with the float tip, in the direction required (see accompanying diagram).

11 Exhaust system

1   Unlike a two-stroke, the exhaust system does not require such frequent attention because the exhaust gases are usually of a less oily nature.
2   Do not run the machine with the exhaust baffles removed, or with a quite different type of silencer fitted. The standard production silencers have been designed to give the best possible performance, whilst subduing the exhaust note to an acceptable level. Although a modified exhaust system, or one without baffles, may give the illusion of greater speed as a result of the changed exhaust note, the chances are that performance will have suffered accordingly.

12 Air cleaner: dismantling and cleaning

1   The air filter box is fitted below the dualseat. Air enters through a vent at the rear of the box containing the element and then passes through the element to ducts which lead to the carburettor mouths. A rubber pipe runs from the breather union on the cylinder head to a union on the lower front edge of the filter box. This is fitted so that all carbon-based vapours expelled from the engine are recirculated through the carburettors and then burnt in the cylinders.
2   To gain access to the filter element the dualseat must first be removed. The seat is secured by two spring loaded catches below and towards the rear of the seat pan.
3   Remove the air filter box lid, which is secured by three screws, and lift out the element retaining frame, followed by the element. The frame is secured at the lower edge by two lugs projecting from the wall of the chamber.
4   The element is of synthetic foam sheeting impregnated with oil and should be cleaned thoroughly in a high-flashpoint solvent. Petrol may be used as an alternative, but the element **must** be allowed to dry completely before the application of new oil. A wet element will constitute a fire risk if blow-back from the carburettors occurs.

12.4 Air filter element is a foam sheet

5   When the element is dry, it should be re-impregnated with clean gear oil (SAE 80 or 90) and squeezed to remove the excess. **Do not** wring out the element because this will cause damage to the fabric necessitating early renewal.

6   If the element is damaged or has become hardened or perished, it should be renewed.

7   On no account run without the air cleaners attached, or with the element missing. The jetting of the carburettors takes into account the presence of the air cleaner and engine performance wili be seriously affected if this balance is upset.

8   To replace the element, reverse the dismantling procedure. Give a visual check to ensure that the inlet hoses are correctly located and not kinked, split or otherwise damaged. Check that the air cleaner cases are free from splits or cracks.

13 Engine lubrication

1   The oil contained in the sump and is shared by the engine, transmission and primary drive.

2   A rotating vane type oil pump driven from a gear on the clutch, delivers oil from the sump to the rest of the engine. Oil is picked up in the sump through a wire gauze which protects the oil pump from any large particles of foreign matter. The delivery section of the pump feeds oil at a preset pressure via a pressure release valve, which by-passes oil to the sump if the pressure exceeds the preset limit. As a result, it is possible to maintain a constant pressure in the lubrication system. The standard pressure is 4.0 – 5.3 kg/cm^2 (56.9 – 75.4 psi) at normal working temperature.

3   Since the oil flow will not, under normal circumstances, actuate the pressure release valve, it passes directly through the full flow filter which has a replaceable element, to filter out any impurities which may otherwise pass to the working parts of the engine. The oil filter unit has its own by-pass valve to prevent the cut-off of the oil supply if the filter element has become clogged.

4   Oil from the filter is fed direct to the crankshaft and big end bearings with a separate pressure feed to the camshaft and valve assemblies. Oil is also fed directly to the balance weight assemblies. Surplus oil drains to the sump where it is picked up again and the cycle is repeated.

5   A fourth oil route is to the transmission through a passage at one end of the mainshaft bearings. After lubrication, the oil is thrown down into the bottom of the crankcase.

6   An oil pressure warning light is included in the lubrication circuit to give visual warning, by means of an indicator light, if the pressure should fall to a low level.

14 Oil pump: dismantling, examination and reassembly

1   The oil pump can be removed from the engine while the engine is still in the frame. However, it will be necessary to remove the primary drive cover to gain access after draining the oil.

2   Place the machine in top gear and apply the rear brake fully. This will prevent rotation of the crankshaft and so allow the oil pump drive sprocket bolt to be released. Remove the bolt and also the circlip which holds the driven sprocket to the oil pump shaft. Pull off both sprockets simultaneously, still meshed with the chain.

3   Loosen the screws holding the oil pump in place and remove the pump and gasket. Displace the drive pin from the pump shaft. This too should be stored safely.

4   Remove the two screws from the pump body and lift off the pump cover plate. Note and remove the shaft end-float shim. Lift the driveshaft and driving pin from position. The inner rotor can now be displaced, followed by the outer rotor.

5   Wash all the pump components with petrol and allow them to dry before carrying out an examination. Before partially reassembling the pump for various measurements to be carried out, check the casing for breakage or fracture, or scoring on the inside perimeter.

6   Reassemble the pump rotors and measure the clearance between the outer rotor and the pump body, using a feeler gauge. If the measurement exceeds the service limit of 0.35 mm (0.014 in) the rotor or the body must be renewed, whichever is worn. Measure the clearance between the outer rotor and the inner rotor, using a feeler gauge. If the clearance exceeds 0.10 mm (0.004 in) the rotors must be renewed as a set. It should be noted that one face of the outer rotor is punch marked. The punch mark should face away from the main pump casing during measurements and on reassembly. With the pump rotors installed in the pump body lay a straight edge across the mating surface of the pump body. Again with a feeler gauge measure the clearance between the rotor faces and the straight edge. If the clearance exceeds 0.10 mm (0.004 in) the rotors should be renewed as a set.

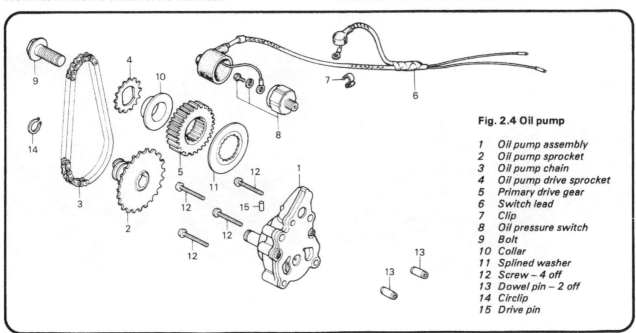

Fig. 2.4 Oil pump

1   Oil pump assembly
2   Oil pump sprocket
3   Oil pump chain
4   Oil pump drive sprocket
5   Primary drive gear
6   Switch lead
7   Clip
8   Oil pressure switch
9   Bolt
10 Collar
11 Splined washer
12 Screw – 4 off
13 Dowel pin – 2 off
14 Circlip
15 Drive pin

7   Examine the rotors and the pump body for signs of scoring, chipping or other surface damage which will occur if metallic particles find their way into the oil pump assembly. Renewal of the affected parts is the only remedy under these circumstances, bearing in mind that the rotors must always be renewed as a matched pair.

8   Reassemble the pump components by reversing the dismantling procedure. Remember that the punch marked face of the rotor must face away from the main pump body. The component parts must be ABSOLUTELY clean or damage to the pump will result. Replace the rotors and lubricate them thoroughly, before refitting the cover. Refit the two hollow dowels and the shim before replacing the cover plate and tightening the screws.

9   Place a new gasket on the two dowels and position the oil pump against the chamber wall. Fit and tighten the screws evenly. Rotate the oil pump gear whilst tightening the screws to check that the rotors do not bind. A stiff pump is almost invariably due to dirty rotors.

14.3 Remove the oil pump

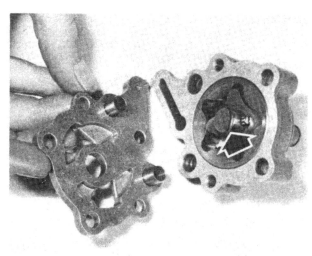

14.4a Separate the halves of the pump. Note the endfloat shim (arrowed)

14.4b Withdraw the driveshaft and pin

14.4c Lift out the inner rotor and then ...

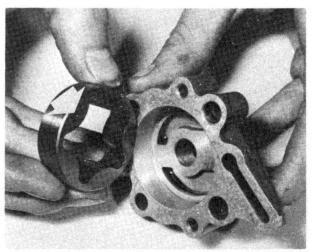

14.4d ... the outer rotor. Note the punch mark (arrowed)

14.6a Measure the clearance between the outer rotor and pump body

14.6b Measure the clearance between the outer rotor and inner rotor

15 Oil pressure by-pass valve: removal, examination and replacement

1   The pressure by-pass valve should be removed periodically to check for correct functioning. The valve consists of a hexagonal body containing a plunger and spring. On early models the valve screws into the crankcase directly below the oil pump and on later models it is mounted inside the crankcase lower half.

2   To dismantle the valve, depress the end cap below the circlip, using a suitable screwdriver, so that the spring pressure is held in check. Remove the circlip and release the pressure in a controlled manner. If the circlip is removed without depressing the cap, the circlip, cap and spring will fly out forcibly, risking the loss of all these components and possibly the sight in one eye.

3   The valve body may now be unscrewed from the casing. Removal is quite possible before the valve components are displaced, but the circlip removal is made easier with the body held firmly in the casing.

4   Clean the valve components thoroughly in petrol and then inspect each one for damage. The valve components cannot be acquired as separate items; if damage to one is evident, the complete assembly must be renewed.

5   Refit the valve by reversing the dismantling procedure.

15.1 Oil pressure by-pass valve directly below oil pump

16 Oil filter: renewing the element

1   The oil filter is contained within a semi-isolated chamber within the crankcase. Access to the element is made by unscrewing the filter cover centre bolt, which will bring with it the cover and also the element. Before removing the cover, place a receptacle beneath the engine, to catch the engine oil contained in the filter chamber.

2   When renewing the filter element it is wise to renew the filter cover O-ring at the same time. This will obviate the possibility of any oil leaks. Do not overtighten the centre bolt on replacement; the correct torque settings is 2.8 – 3.2 kgf m (20 – 23 lbf ft).

3   The filter by-pass valve, comprising a plunger and spring, is situated in the bore of the filter cover centre bolt. It is recommended that the by-pass valve be checked for free movement during every filter change. The spring and plunger are retained by a pin across the centre bolt. Knocking the pin out will allow the spring and plunger to be removed for cleaning.

4   Never run the engine without the filter element or increase the period between the recommended oil changes or oil filter changes. Engine oil should be changed every 2000 miles and the element changed every 4000 miles. Use only the recommended viscosity of oil.

17 Oil pressure warning lamp

1   An oil pressure warning lamp is incorporated in the lubrication system to give immediate warning of excessively low oil pressure.

2   The oil pressure switch is screwed into the crankcase in a horizontal position, protected by a shield incorporated in the primary drive cover. The switch is interconnected with a warning light on the lighting panel on the handlebars. The light should be on whenever the ignition is on but will usually go out at about 1500 rpm.

3   If the oil warning lamp comes on whilst the machine is being ridden, the engine should be switched off immediately, otherwise there is a risk of severe engine damage due to lubrication failure. The fault must be located and rectified before the engine is re-started and run, even for a brief moment. Machines fitted with plain shell bearings rely on high oil pressure to maintain a thin oil film between the bearing surfaces. Failure of the oil pressure will cause the working surfaces to come into direct contact, causing overheating and eventual seizure.

18 Fault diagnosis: fuel system and lubrication

Symptom	Cause	Remedy
Engine gradually fades and stops	Fuel starvation Sediment in filter bowl or float chamber	Check vent hole in filler cap. Dismantle and clean.
Engine runs badly. Black smoke from exhausts	Carburettor flooding	Dismantle and clean carburettor. Check for punctured float or sticking float needle.
Engine lacks response and overheats	Weak mixture Air cleaner disconnected or hose split Modified silencer has upset carburation	Check for partial block in carburettors. Reconnect or renew hose. Replace with original design.
Oil pressure warning light comes on	Lubrication system failure	Stop engine immediately. Trace and rectify fault before re-starting.
Engine gets noisy	Failure to change engine oil when recommended	Drain off old oil and refill with new oil of correct grade. Renew oil filter element.

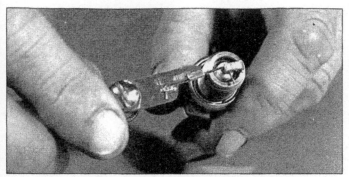

Electrode gap check - use a wire type gauge for best results

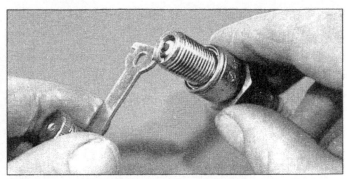

Electrode gap adjustment - bend the side electrode using the correct tool

Normal condition - A brown, tan or grey firing end indicates that the engine is in good condition and that the plug type is correct

Ash deposits - Light brown deposits encrusted on the electrodes and insulator, leading to misfire and hesitation. Caused by excessive amounts of oil in the combustion chamber or poor quality fuel/oil

Carbon fouling - Dry, black sooty deposits leading to misfire and weak spark. Caused by an over-rich fuel/air mixture, faulty choke operation or blocked air filter

Oil fouling - Wet oily deposits leading to misfire and weak spark. Caused by oil leakage past piston rings or valve guides (4-stroke engine), or excess lubricant (2-stroke engine)

Overheating - A blistered white insulator and glazed electrodes. Caused by ignition system fault, incorrect fuel, or cooling system fault

Worn plug - Worn electrodes will cause poor starting in damp or cold weather and will also waste fuel

Chapter 3 Ignition system

Contents

Specifications

Alternator
Make .... Nippon Denso (250 model), Hitachi (400 model)
Type .... Rotating permanent magnet, fixed multi-coil stator

Ignition
Type .... Capacitor discharge ignition (CDI)

Ignition timing

	250N	400N
Retarded:		
On 'F' mark	15° BTDC @ 1300 rpm	15° BTDC @ 1200 rpm
Full advance	47° BTDC @ 5450–6300 rpm	43° BTDC @ 4500–5350 rpm

Sparking plugs

	NGK*	ND*	Motorcraft
Make	DR8ES	X27ESR-U*	HG2
Type			
*Manufacturer's recommendation			
Plug gap	0.6–0.7 mm (0.024–0.028 in)		

1 General description

The ignition system used on all the models covered in this manual is of the CDI (capacitive discharge ignition) type. In this type of system the normal contact breaker and the mechanical automatic timing unit are dispensed with, their place being taken by solid-state components which control the ignition timing electronically. The motorcycle's electrical system is powered by a crankshaft mounted 12 volt alternator, which also incorporates the ignition timing pick-up and the pulser unit which controls the ignition timing automatic advance. Mounted close to the alternator is the fixed pulser unit, which triggers the current to the CDI unit mounted below the fuel tank.

2 CDI system: principles of operation

1 As the alternator rotor rotates, alternating current (ac) is generated in the ignition source coil of the stator. This current is then rectified by a diode and stored in the capacitor (condenser). When the pick-up, which is mounted on the periphery of the rotor, passes the fixed pulser, a signal current is induced in the pulser which is applied to an electrical switch. The switch is turned on, allowing the current stored in the condenser to discharge through the ignition primary coil. This in turn produces a high current in the ignition secondary coil, which is passed through the HT leads to the sparking plugs. The current running to earth across the sparking plug electrodes arcs, causing combustion to take place.

2 The ignition timing advance is controlled automatically, by the signal current from the fixed pulser being replaced by one from the advance pulser contained within the alternator. As engine speed increases, the advance pulser allows the electrical switch to be operated earlier and the current to be passed through the primary windings in the usual way.

3 Checking the ignition source coil

1 The performance and output of both the source coils for the charging circuit and the ignition source coil, can only be efficiently checked with specialised test equipment of the multimeter type. It is unlikely that the average owner/rider will have access to this type of equipment or the instruction in its use. In consequence, if the performance is suspect it should be checked by a Honda Service Agent or auto-electrician.

2 Diminished performance or complete failure of the charging system may be caused by faults other than in the coil windings. Check that the coil mounting bolts are tight, as poor earthing results in low or erratic output, and check that all main wiring connections are firm and bright. Total output failure on all the coils may be due to the flywheel having sheared its drive key on the crankshaft drive taper. This fault will be self-evident, on inspection.

2.1 Capacitor discharge ignition pick-up (arrowed)

4 CDI unit and fixed pulser: testing

1 The comments in the preceding Section relating to the testing of suspect components applies equally to the CDI unit or the pulser unit. In the event of failure, the part in question should be renewed.

5 Ignition timing: checking

1 Because no mechanical means is used to control the point at which ignition occurs, ignition timing can be checked only when the engine is running, using a stroboscopic lamp.

2 Remove the alternator cover from the left-hand side of the engine, so that the alternator rotor is in view. Note the various timing marks on the rotor periphery and the index mark projecting from the top of the casing. Connect the stroboscope with the ignition by following the instructions given by the lamp's manufacturer.

3 Start the engine and allow it to run at tick-over speed. Aim the beam of the stroboscopic lamp at the index mark. If at the correct speed, the 'F' mark aligns with the index mark, the retarded ignition timing is correct.

4 To check the performance of the automatic advance system, increase the engine speed progressively until the index mark is between the two adjacent full advance marks on the rotor periphery. If the advance progression does not take place, or if full advance is reached outside the specified engine speed range, the advance unit is not functioning correctly.

5 The ignition timing is corrected and set permanently during manufacture, no adustment being possible at a later date. If the ignition timing is found to be incorrect, or if the advance system malfunctions, the only remedy is renewal of the parts concerned.

6 Sparking plugs: checking and resetting the gaps

1 The two NGK or Nippon-Denso sparking plugs are fitted to the CB250N and CB400N models as standard. The plug types are given in the specifications at the beginning of the Chapter. Certain operating conditions may indicate a change in sparking plug grade, but generally the type recommended by the manufacturer gives the best all round service.

2 Check the gap of the plug points at the three monthly or 3600 mile service. To reset the gap, bend the outer electrode to bring it closer to, or further away from, the central electrode until a 0.7 mm (0.028 in) feeler gauge can be inserted. Never bend the centre electrode or the insulator will crack, causing engine damage if the particles fall into the cylinder whilst the engine is running.

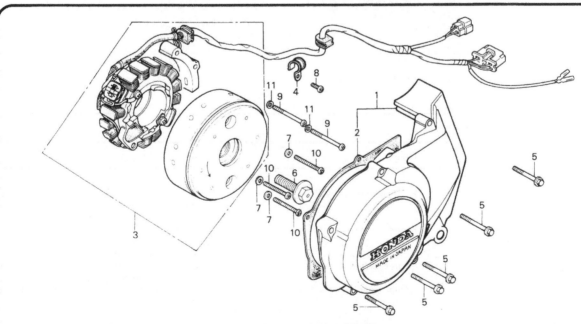

Fig. 3.1 Alternator assembly

1 LH crankcase cover	5 Bolt – 5 off	9 Screw – 2 off
2 Gasket	6 Bolt	10 Screw – 3 off
3 Alternator – complete	7 Washer – 3 off (400N model)	11 Washer – 2 off (250N model)
4 Clip	8 Screw	

3   With some experience, the condition of the sparking plug electrodes and insulator can be used as a reliable guide to engine operating conditions.

4   Always carry spare sparking plugs of the recommended grade. In the rare event of plug failure, this will enable the engine to be restarted.

5   Beware of over-tightening the sparking plugs, otherwise there is risk of stripping the threads from the aluminium alloy cylinder heads. The plugs should be sufficiently tight to seat firmly on their copper sealing washers, and no more. Use a spanner which is a good fit to prevent the spanner from slipping and breaking the insulator.

6   If the threads in the cylinder head strip as a result of over-tightening the sparking plugs, it is possible to reclaim the head by the use of a Helicoil thread insert. This is a cheap and convenient method of replacing the threads; most motorcycle dealers operate a service of this nature at an economic price.

7   Make sure the plug insulating caps are a good fit and have their rubber seals. They should also be kept clean to prevent tracking. These caps contain the suppressors that eliminate both radio and TV interference.

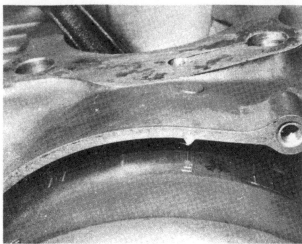

5.3a The retarded timing mark 'F'

5.3b The full advance timing marks

7   Fault diagnosis: ignition system

Symptom	Cause	Remedy
Engine will not start	Faulty ignition switch	Operate switch several times in case contacts are dirty. If lights and other electrics function, switch may need renewal.
	Starter motor not working	Discharged battery. Use kickstart until battery is recharged. Faulty starter inhibitor switch.
	Short circuit in wiring	Check whether fuse is intact. Eliminate fault before switching on again.
	Completely discharged battery	If lights do not work, remove battery and recharge.
Engine misfires	Fouled sparking plug	Renew plug and have original cleaned.
	Poor sparking due to generator failure and discharged battery	Check output from generator. Remove and recharge battery.
Engine lacks power and overheats	Retarded ignition timing	Check timing.
Engine 'fades' when under load	Pre-ignition	Check grade of plugs fitted; use recommended grades only.

Chapter 4 Frame and forks

Contents

Specifications

Frame

Type	Diamond, using engine as a stressed member

Front forks

Type	Telescopic, hydraulically damped
Travel	139·5 mm (5·5 in)
Spring free length (min)	480 mm (18·90 in)
Oil capacity (per leg)	140 cc (4.9 fl oz)

Rear suspension

Type	Swinging arm, supported on two hydraulically damped suspension units
Travel	96 mm (3·8 in)

1 General description

Both models covered by this manual share a frame design of similar type, where the engine is a stressed member forming a structural part of the frame. By adopting this style of construction, weight can be reduced and engine removal becomes easier.

The front forks are of the conventional telescopic type, having internal, oil-filled dampers. The fork springs are contained within the fork stanchions and each fork leg can be detached from the machine as a complete unit, without dismantling the steering head assembly.

Rear suspension is of the swinging arm type, using oil filled suspension units to provide the necessary damping action. The units are adjustable so that the spring ratings can be effectively changed within certain limits, to match the load carried.

2 Front fork legs: removal from the frame

1 It is unlikely that the front forks will have to be removed from the frame as a complete unit, unless the steering head assembly requires attention or if the machine suffers frontal damage. If this is the case, remove the fork legs by following the procedure in this Section and then continue dismantling as described in the following Section.
2 Place a sturdy support below the crankcase so that the front wheel is raised well clear of the ground. Remove the speedometer cable from the gearbox on the left-hand side of the wheel; the cable is held by a screw.
3 Remove the two bolts which retain the brake caliper assembly on the single disc of the 250N, and the four bolts which retain the two caliper assemblies on the dual discs of the 400N. Remove the assembly together with the associate brake

operating components. Each brake hose is held by a clip and grommet secured by a single bolt. Free the clip and pull the caliper and hose backwards, away from the forks. This is to be done with both calipers on the 400N. Tie each caliper to a suitable point on the frame, clear of any parts to be removed.

4   Remove the split pin and undo the castellated nut on the wheel spindle. Loosen the bolts which hold the spindle clamp on the right-hand fork lower leg. Pull the wheel spindle from position and allow the front wheel to drop free.

5   When removing the various hydraulic brake parts it is important that care is taken not to bend or kink the brake hoses or pipes. Under no circumstances should the front brake be operated at any time during removal or there is danger of the caliper piston being forced out of the cylinder, with the resulting loss of fluid. Should this happen, the brake assembly must be bled after it has been replaced and the front wheel is in position. If any fluids is inadvertently spilled onto the paintwork it should be removed at once. Hydraulic fluids are most effective paint removers.

6   Remove the front mudguard which is retained by two bolts through brackets on the inside of each fork lower leg. Also remove the metal guide on the left inside of the mudguard (CB250N) and on both sides of the 400 model. Remove the mudguard complete with centre bracket.

7   Unscrew the large chrome cap from each fork stanchion (upper tube) top, and then slacken the pinch bolts clamping each fork leg to the lower steering yoke. Each fork leg can now be removed as a complete unit. It may be necessary to spring the yoke clamps apart with a screwdriver to allow the fork legs to be pulled down, out of position.

8   The fork legs may now be dismantled individually as described in Section 4.

3   Fork yoke and steering head bearings: removal

1   The preceding Section describes the procedure for removal of the fork legs without having to disturb the steering head bearings or yokes. Further dismantling of the fork assembly may be accomplished as follows, after carrying out fork leg removal.

2   Commence by removing the right-hand side cover from the

frame and disconnecting the accessible battery lead from the terminal. Isolating the electrical system in this way will prevent accidental short-circuits when the wiring leads are disconnected, during dismantling.

3   Release the headlamp/rim reflector assembly by removing the two screws that pass through the shell at the 8 o'clock and 4 o'clock positions and into the assembly. Disconnect the bulb leads and place the rim/reflector unit to one side.

4   Detach the main controls from the handlebars. The hydraulic master cylinder must be removed. The master cylinder and operating lever assembly is clamped to the bars by two bolts and nuts. Hold the master cylinder in an upright position to prevent fluid spillage. In the case of the 250N the cylinder and hose can now be threaded across the machine so that they and the caliper may be removed from the machine as an undismantled assembly. On the 400N model there are two interconnected hoses which must be removed, together with two calipers, as a complete assembly.

5   Disconnect the main connections within the headlamp shell at the block connectors and snap connectors. Similarly disconnect the leads from the speedometer head, tachometer and the ignition switch. The wires leading through to the handlebars can be pulled through the orifice in the rear of the headlamp, and the handlebars switches completely removed.

6   Slacken the four bolts that retain the handlebar clamps, and lift the handlebars away complete with tachometer and speedometer which are retained on a common bracket bolted to the top steering yoke. Disconnect the two drive cables by unscrewing the knurled ring on each cable end and the indicator lamps.

7   Remove the large domed nut from the top of the steering stem and, holding the headlamp shell steady, remove the fork upper yoke. The headlamp can be lifted off the lower yoke complete with the two fork shroud/headlamp brackets.

8   To release the lower yoke and the steering head stem, unscrew the adjuster ring at the top with a suitable C-spanner. If such a spanner is not available, a brass drift can be used to loosen the ring. As the steering head is lowered, the uncaged ball bearings from the lower race will be released, and care should be taken to catch them as they fall free. The bearings in the upper race will almost certainly stay in place. These can be dislodged once the lower yoke is out of the way.

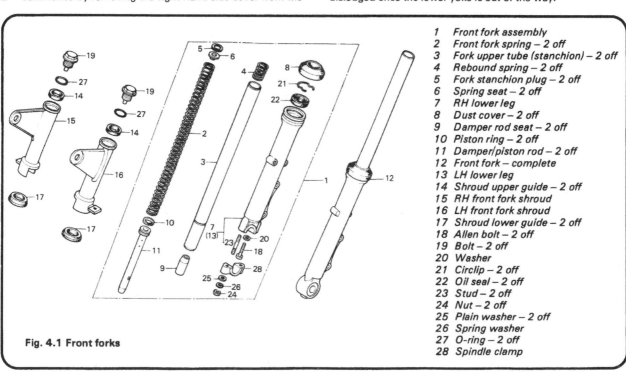

Fig. 4.1 Front forks

1   Front fork assembly
2   Front fork spring – 2 off
3   Fork upper tube (stanchion) – 2 off
4   Rebound spring – 2 off
5   Fork stanchion plug – 2 off
6   Spring seat – 2 off
7   RH lower leg
8   Dust cover – 2 off
9   Damper rod seat – 2 off
10 Piston ring – 2 off
11 Damper/piston rod – 2 off
12 Front fork – complete
13 LH lower leg
14 Shroud upper guide – 2 off
15 RH front fork shroud
16 LH front fork shroud
17 Shroud lower guide – 2 off
18 Allen bolt – 2 off
19 Bolt – 2 off
20 Washer
21 Circlip – 2 off
22 Oil seal – 2 off
23 Stud – 2 off
24 Nut – 2 off
25 Plain washer – 2 off
26 Spring washer
27 O-ring – 2 off
28 Spindle clamp

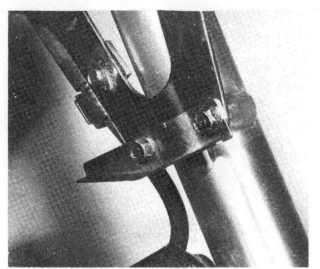

2.6 The front mudguard is held by two bolts on each leg

2.7a Remove the large chrome bolts and ...

2.7b ... slacken the yoke pinch bolts

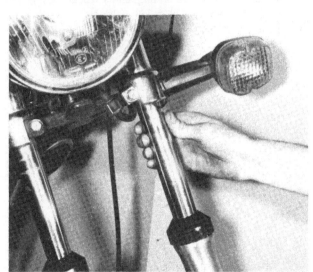

2.7c The fork legs can be withdrawn from the yokes

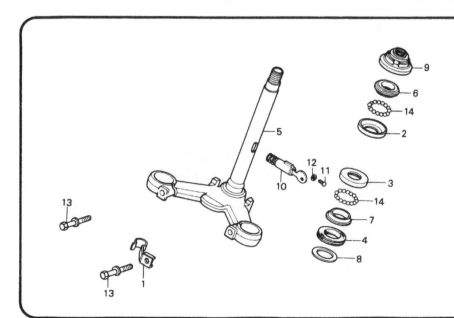

Fig. 4.2 Steering head bearings

1   Hose clip (not 400N)
2   Upper cup
3   Lower cup
4   Dust seal
5   Lower yoke/steering stem
6   Upper cone
7   Lower cone
8   Washer
9   Bearing adjuster nut
10 Lock
11 Rivet
12 Washer
13 Special bolt – 2 off
14 Steel ball

4   Front forks: dismantling the fork legs

1   It is advisable to dismantle each fork leg separately using an identical procedure. There is less chance of unwittingly exchanging parts if this approach is adopted.

2   Grasp the top of the fork leg stanchion in the jaws of a vice so that the leg is upright. Protect the stanchion surface by wrapping a heavy cloth or length of inner tubing around the gripped portion. Do not overtighten the vice. Using a socket key, unscrew the inner plug that retains the fork spring. When the plug is approaching the end of the threads, apply downward pressure to control the pressure in the fork spring. Release the plug and lift out the fork spring. Note that the spring will be covered in oil because the oil has yet to be drained.

3   Remove the fork leg from the vice and drain out the oil into a suitable container. Draining is aided if the leg is pumped a number of times.

4   Relocate the fork leg in the vice so that the lower leg is grasped, and remove the Allen screw from the recess in the base of the leg. Prise the dust cover off the lower fork leg and pull the fork tube out of the lower leg. If the fork tube is now inverted, the damper rod assembly will fall out. Fitted to the damper rod is a single piston ring: and a rebound spring. The ring need not be removed, except for renewal.

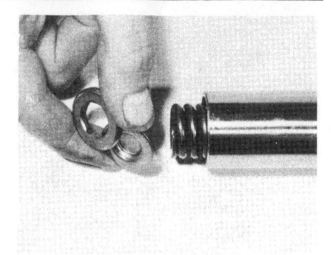

4.2a Unscrew the inner plug retaining the fork spring

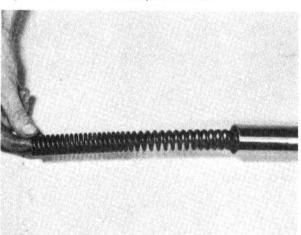

4.2b Remove the fork spring from the stanchion

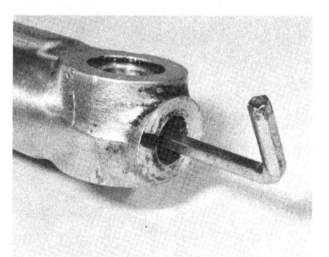

4.4a Unscrew the Allen bolt ...

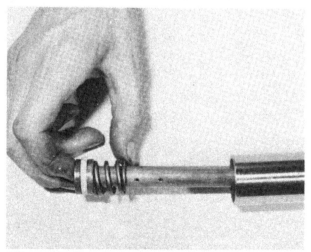

4.4b ... and pull out the damper rod assembly

4.4c The damper rod assembly

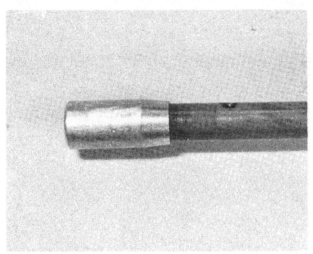

4.4d The damper rod seat

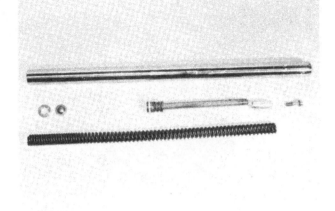

4.4e The dismantled front fork assembly complete

5 Front forks: examination and renovation

1   The parts most liable to wear over an extended period of service are the internal surfaces of the lower leg and the outer surfaces of the fork stanchion or tube. If there is excessive play between these two parts, they must be renewed as a complete unit. Check the fork tube for scoring over the length which enters the oil seal. Bad scoring here will damage the oil seal and lead to fluid leakage.

2   It is advisable to renew the oil seals when the forks are dismantled, even if they appear to be in good condition. This will save a strip down of the forks at a later date if oil leakage occurs. The oil seal in the top of each lower fork leg is retained by an internal C-ring which can be prised out of position with a small screwdriver. Check that the dust excluder rubbers are not split or worn where they bear on the fork tube. A worn excluder will allow the ingress of dust and water which will damage the oil seal and eventually cause wear of the fork tube.

3   It is not generally possible to straighten forks which have been badly damaged in an accident, particularly when the correct jigs are not available. It is always best to err on the side of safety and fit new ones, especially since there is no easy means to detect whether the forks have been over stressed or metal fatigued. Fork stanchions (tubes) can be checked, after removal from the lower legs by rolling them on a dead flat surface. Any misalignment will be immediately obvious.

4   The fork springs will take a permanent set after considerable usage and will need renewal if the fork action becomes spongy. The service limit for the total free length of each spring is 480 mm (18.90 in). Always renew them as a matched pair.

6 Steering head bearings: examination and renovation

1   Before commencing reassembly of the forks, examine the steering head races. The ball bearing tracks of the respective cup and cone bearings should be polished and free from indentations, cracks or pitting. If signs of wear are evident, the cups and cones must be renewed. In order for the straight line steering on any motorcycle to be consistently good, the steering head bearings must be absolutely perfect. Even the smallest amount of wear on the cups and cones may cause steering wobble at high speeds and judder during heavy front wheel braking. The cups and cones are an interference fit on their respective seatings and can be tapped from position with a suitable drift.

2   Ball bearings are relatively cheap. If the originals are marked or discoloured they must be renewed. To hold the steel balls in place during reassembly of the fork yokes, pack the bearings with grease. The upper race is fitted with eighteen $\frac{1}{4}$ inch steel balls and the lower race nineteen balls. Although each race has room for an extra steel ball it must not be fitted. The gap allows the bearings to work correctly, stopping them skidding and accelerating the rate of wear.

7 Front forks: replacement

1   Replace the front forks by following in reverse the dismantling procedures described in Sections 2 and 4 of this Chapter. Each fork leg should be filled with 140 cc (4.9 fl oz) of good quality automatic transmission fluid (ATF) before the screw plug is fitted into the top of the stanchion.

2   If the fork stanchions prove difficult to re-locate through the fork yokes, make sure their outer surfaces are clean and polished so that they will will slide more easily. It is often advantageous to use a screwdriver blade to open up the clamps as the stanchions are pushed upwards into position. Ensure that the fork legs are fitted correctly, with the rim of each stanchion flush with the top of the yoke eyes. With the stanchions so placed, tighten the lower yoke pinch bolts. Before fully tightening the front wheel spindle clamps bounce the forks several times to ensure they work freely and are clamped in their original positions.

3   Before the machine is used on the road, check the adjustment of the steering head bearings. If they are too slack, judder will occur. There should be no detectable play in the head races when the handlebars are pulled and pushed with the front brake applied hard.

4   Overtight head races are equally undesirable. It is possible to unwittingly apply a loading of several tons on the head bearings by overtightening, even though the handlebars appear to turn quite freely. Overtight bearings will cause the machine to roll at low speeds and give generally imprecise handling, with a tendency to weave. Adjustment is correct if there is no perceptible play in the bearings and the handlebars will swing to full lock in either direction, when the machine is on the centre stand with the front wheel clear of the ground. Only a slight tap should cause the handlebars to swing.

5   The adjustment should be made using a C-spanner applied to the special nut on the steering stem, immediately below the instrument mounting plate. Make the adjustment with the domed nut on the top of the steering stem loose.

5.2 The oil seal is held by a spring clip

7.1 Replenish each fork leg with the correct quantity of oil

8 Steering head lock

1   The steering head lock is fitted in the underside of the lower yoke of the forks, where it is secured by a cover and rivet. When in the locked position, a bolt extends from the body of the lock when the forks are on full left-hand lock and abut against a portion of the steering head lug.
2   If the lock malfunctions it must be renewed. A repair is impracticable. When the lock is changed a new key must of course be used to match the new lock.
3   To remove the lock, dislodge the rivet and lift off the cover. Insert the key and twist it to the left. The key may be then withdrawn together with the lock unit.

8.1 The steering head lock fitted below the steering head

9   Frame: examination and renovation

1   The frame is unlikely to require attention unless accident damage has occurred. In some cases, replacement of the frame is the only satisfactory course of action if it is badly out of alignment. Only a few frame repair specialists have the jigs and mandrels necessary for resetting the frame to the required standard of accuracy and even then there is no easy means of

assessing to what extent the frame may have been over-stressed.
2   After the machine has covered a considerable mileage, it is advisable to examine the frame closely for signs of cracking or splitting at the welded joints. Rust can also cause weakness at these joints. Minor damage can be repaired by welding or brazing, depending on the extent and nature of the damage.
3   Remember that a frame which is out of alignment will cause handling problems and may even promote 'speed wobbles'. If misalignment is suspected, as the result of an accident, it will be necessary to strip the machine completely so that the frame can be checked and, if necessary, renewed.

10 Swinging arm rear fork: dismantling, examination and renovation

1   The rear fork assembly pivots on shouldered bushes pressed into the cross-member at the end of each fork arm, which are supported by a pivot shaft running through the frame down tube box-section. Worn swinging arm pivot bearings will give imprecise handling, with a tendency for the rear end of the machine to twitch or hop. The play can be detected by placing the machine on its centre stand and with the rear wheel clear of the ground, pulling and pushing on the fork ends in a horizontal direction. Any play will be magnified by the leverage effect. In the UK, excess play will cause the machine to fail the DOT test. It is quite easy to renovate the swinging arm when wear necessitates attention.
2   To remove the rear swinging arm fork, first position the machine on the centre stand so that it rests firmly and securely. Remove the final drive chain by detaching the master link, and then unscrew the two bolts that hold the chain guard.
3   Detach the brake torque arm from the lug on the brake plate and from the lug on the frame. In both cases the torque arm bolt retaining nuts are secured by split pins. Unscrew the adjuster nut on the brake operating arm and pull the rod through the trunnion on the brake arm. Replace the adjuster nut to avoid the loss of brake rod spring.
4   Remove the split pin from the end of the wheel spindle and remove the castellated nut. The wheel spindle can now be pulled out to the left. Knock the wheel spacer out of position from between the brake plate and the fork end. The rear wheel can now be lifted rearwards, out of the frame.
5   Remove the lower bolt from each rear suspension unit and swing the units back, out of the way.

6   Remove the locknut from the end of the pivot shaft, which can then be tapped out from the left-hand side. Working the swinging arm fork up and down will aid removal of the shaft. The swinging arm fork is now free to be pulled from position between the two frame lugs.

7   Remove the dust excluder caps from each end of the fork crossmember. Note the presence of the lipped sealing rings in the caps. Push out the two inner bushes.

8   Wash the inner and outer bushes carefully in petrol or another solvent. Do not remove the outer bushes from position in the fork crossmember unless they need renewal as they are made of a brittle material that will probably fracture while being drifted out. When drifting in new bushes, ensure that they enter their housings squarely. Use a soft wood or hard rubber pad between the bush and drive, to prevent chipping. Check the pivot shaft for straightness by rolling it on the edge of a dead flat surface. If the shaft is bent it must be renewed or straightened.

9   Reassemble the swinging arm fork by reversing the dismantling procedure. Grease the pivot shaft and bearings liberally before reassembly and check that the sealing rings in

the dust caps are in good condition. After installation, grease the bearings thoroughly by applying a grease gun to the nipple provided. Clean off any excess grease that has exuded from the ends of the cross-members.

11 Rear suspension units: examination

1   The rear suspension units fitted to the Honda 250N and 400N models are of the normal hydraulically damped type, adjustable to give five different spring settings. A C-spanner included in the tool kit should be used to turn the lower spring seat and so alter its position on the adjustment projection. When the spring seat is turned so that the effective length of the spring is shortened the suspension will become heavier.

2   If a suspension unit leaks, or if the damping efficiency is reduced in any other way, the two units must be renewed as a pair. For precise roadholding, it is imperative that both units react to movement in the same way. It follows that the units must always be set at the same spring loading.

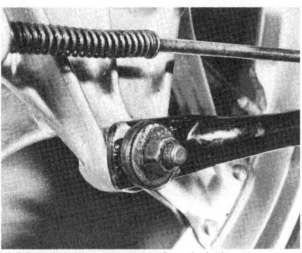

10.3 Detach the brake torque arm from the brake plate

10.4 Pull the wheel spindle out to the left

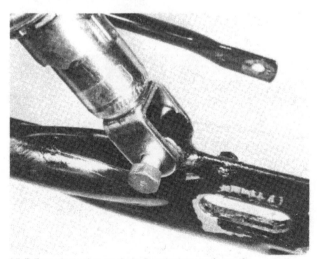

10.5 Detach the lower end of each suspension unit

10.6a Remove the locknut from the pivot shaft end and ...

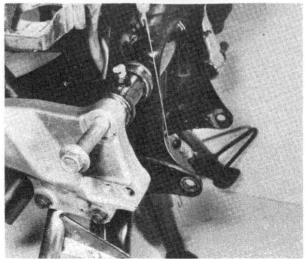
10.6b ... withdraw the swinging arm pivot shaft

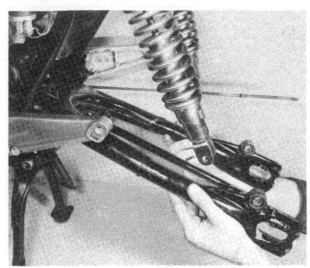
10.6c The swinging arm fork can be pulled clear

10.7a Pull off the dust covers and ...

10.7b ... push out the inner bushes

10.7c The shouldered bushes may be driven out

10.9 After refitting, grease the bushes thoroughly

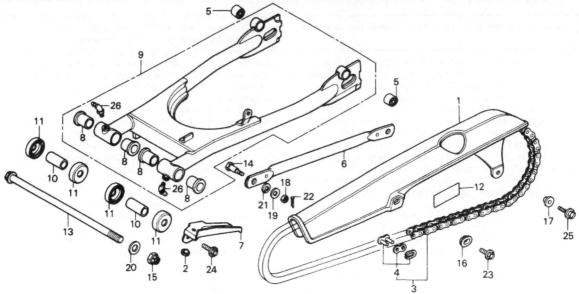

Fig. 4.3 Rear swinging arm fork

1 Chain case	8 Fork bush – 4 off	15 Locking nut	22 Split pin
2 Lower guard stop	9 Swinging arm fork	16 Special washer	23 Bolt
3 Final drive chain	10 Spacer – 2 off	17 Special washer	24 Bolt
4 Master link	11 Dust cover – 4 off	18 Nut	25 Bolt
5 Bonded rubber bush – 2 off	12 Instruction label	19 Plain washer	26 Grease nipple – 2 off
6 Rear brake torque arm	13 Swinging arm pivot	20 Plain washer	
7 Lower guard	14 Shouldered bolt	21 Spring washer	

12 Centre stand: examination

1 The centre stand pivots on a shaft running across the bottom frame tubes, which is secured by a pinch bolt at the right-hand end. The pivot assemblies on centre stands are often neglected with regard to lubrication and this will eventually lead to wear. It is prudent to remove the pivot shaft from time to time and grease it thoroughly. This will prolong the effective life of the stand.

2 Check that the return spring is in good condition. A broken or weak spring may cause the stand to fall whilst the machine is being ridden, and catch in some obstacle, unseating the rider.

13 Prop stand: examination

1 The prop stand is attached to a lug welded to the left-hand lower frame tube. A rubber block has been added to the lower end of the prop stand so that it will retract automatically when it strikes an obstacle if it has been left down inadvertently.

2 Check that the pivot bolt is secured and that the extension spring is in good condition and not overstretched. Check the rubber block for wear or damage. No part should be worn below the moulded line on the rubber.

14 Footrests: examination and renovation

1 Both the front footrests and the pillion footrests are bolted to heavy cast aluminium plates, one of which is fitted each side of the frame. The footrests pivot upwards on their mounting brackets and are spring loaded, to keep them in their horizontal position. If an obstacle is struck, they will fold upwards, reducing the risk of damage to the rider's foot or to the main frame.

2 If the footrests are damaged in an accident, it is possible to dismantle the assembly into its component parts. Detach each footrest from the mounting plate and separate the folding foot piece from the bracket on which it pivots by withdrawing the split pin and pulling out the pivot shaft. It is preferable to renew the damaged parts, but if necessary, they can be bent straight by clamping them in a vice and heating to a dull red with a blow lamp whilst the appropriate pressure is applied. Do not attempt to straighten the footrests while they are attached to the frame.

3 If heat is applied to the main footrest piece during any straightening operation it follows that the footrest rubber must be removed temporarily.

15 Rear brake pedal: examination and renovation

1 The rear brake pedal pivots on a bolt which passes through the right-hand footrest plate and screws into the end of the centre stand pivot shaft.

2 If the brake pedal is bent or twisted it can be removed after disconnecting the return spring and the brake rod, and unscrewing the pivot bolt. The pedal should be straightened by adopting the same method as recommended for bent footrests. It should be borne in mind that heating the pedal will almost certainly destroy the chrome plate with which the component is finished, therefore if the cosmetic appearance of the machine is important the part should be renewed.

3 The rear brake pedal is returned to its normal position by an extension spring. This should be checked to ensure that it is not stretched, and pulls the brake off cleanly.

16 Dualseat: removal

1 The dualseat is secured to the frame by a bridge at the front which locates with the frame, and by two spring loaded catches at the rear. A single chain link on the left-hand side of the seat may be engaged with the helmet lock unit to prevent seat removal.

2 Periodically the seat should be removed and the spring loaded catches lubricated with light oil.

17 Speedometer head and tachometer head: removal and replacement

1 The speedometer head and the tachometer head may be removed from the machine individually by removing the instrument console cover and releasing the two nuts holding each instrument, or together by detaching the complete console and then separating the components, after removal. If the former method is adopted, the bulb holders should be removed from the base of the instruments as they are lifted from position.

2 The drive cables must be detached first in either case. Each is secured by a knurled ring. Apart from defects in either the drive or drive cables, a speedometer or tachometer which malfunctions is difficult to repair. Fit a replacement or alternatively entrust the repair to a competent instrument repair specialist.

3 Remember that a speedometer in correct working order is a statutory requirement in the UK. Apart from this legal necessity, reference to the odometer readings is the most satisfactory means of keeping pace with the maintenance schedules.

18 Speedometer and tachometer drive cables: examination and maintenance

1 It is advisable to detach the drive cables from time to time in order to check whether they are lubricated adequately, and whether the outer coverings are damaged or compressed at any point along their run. Jerky or sluggish movements can often be traced to a damaged drive cable.

2 For greasing, withdraw the inner cable. After removing all the old grease, clean with a petrol-soaked rag and examine the cable for broken strands or other damage.

3 Regrease the cable with high melting point grease, taking care not to grease the last six inches at the point where the cable enters the instrument head. If this precaution is not observed, grease will work into the head and immobilise the instrument movement.

4 If any instrument head stops working suspect a broken drive cable unless the odometer readings continue. Inspection will show whether the inner cable has broken; if so, the inner cable alone can be replaced and re-inserted in the outer casing after greasing. Never fit a new inner cable alone if the covering is damaged or compressed at any point along its run.

19 Speedometer and tachometer drives: location and examination

1 The speedometer drive gearbox is fitted on the left-hand side of the front wheel hub. The drive rarely gives trouble provided it is kept properly lubricated. Lubrication should take place whenever the front wheel is removed for wheel bearing inspection or replacement.

2 The tachometer drive is taken from the primary drive cover via a gear on the end of the oil pump driven pinion, and then through a flexible cable to the tachometer head. It is unlikely that the internal drive will give trouble during the normal service life of the machine, particularly since it is fully enclosed and effectively lubricated.

20 Fault diagnosis: frame and forks

Symptom	Cause	Remedy
Machine veers either to the left or the right with hands off handlebars	Bent frame Twisted forks Wheels out of alignment	Check and renew. Check and renew. Check and re-align.
Machine rolls at low speed	Overtight steering head bearings	Slacken until adjustment is correct.
Machine judders when front brake is applied	Slack steering head bearings Worn fork sliders	Tighten until adjustment is correct. Dismantle forks and renew lower legs and/or stanchions.
Machine pitches on uneven surfaces	Ineffective fork dampers Ineffective rear suspension units Suspension too soft	Check oil content. Check whether units still have damping action. Raise suspension unit adjustment one notch.
Fork action stiff	Fork legs out of alignment (twisted in yokes)	Slacken yoke clamps, and fork top bolts. Pump fork several times then retighten from bottom upwards.
Machine wanders. Steering imprecise. Rear wheel tends to hop	Worn swinging arm pivot	Dismantle and renew bushes and pivot shaft.

Chapter 5 Wheels, brakes and tyres

Contents

Specifications

Tyres

	250N	400N
Front	3·60S19	3·60S19
Rear	4·10S18	4·10S18

Tyre pressures

	250N	400N
Front	24 psi	24 psi
*Rear	32 psi	32 psi

*Increase by 4 psi when carrying a passenger or travelling at high speed

Brakes

Front	Hydraulically operated 273 mm (10.75 in) single disc Single leading shoe 140 mm (5·5 in) drum	Hydraulically operated 239 mm (9.4 in) twin disc Single leading shoe 140 mm (5·5 in) drum
Rear		

1 General description

The two models covered in this manual are fitted as standard with an 18 inch diameter rear wheel and a 19 in front wheel. The front tyre section is 3.60 in and the rear 4.10 in. The wheels consist of an aluminium alloy rim and a hub interconnected by aluminium alloy spokes. The components are riveted together to form a unit, which has a potential strength greater than that of the spoked wheel.

The CB250N is fitted with a single disc brake at the front and a single leading shoe drum rear brake. The CB400N utilises the same rear brake but has a twin disc front brake.

2 Front wheel: examination and renovation

1   The Comstar wheel requires less routine attention than a spoked wheel, in that periodic tensioning of the spokes is neither necessary nor possible.
2   Using a fixed pointer, check the wheel rim for misalignment when it is rotated. Honda recommend the following maximum permissible limits:

Radial run-out   2.0 mm (0.08 in)
Axial run-out   2.0 mm (0.08 in)

The wheel cannot be repaired if buckling or damage to the spokes has taken place. In the event of such damage, the wheel must be renewed as a complete unit. Do not attempt to tighten or loosen the bolts which hold the plate spokes to the wheel hub. The nuts are tightened and secured by pins at the factory; and should not be disturbed.

3 Front wheel: removal and replacement

1   Place the machine on the centre stand so that the front wheel is clear of the ground. If necessary, place a block below the front of the engine to balance the machine.
2   The speedometer cable on both models is a push fit at the lower end, secured by a single screw. Unscrew the screw and withdraw the cable.

3   On 400 cc models one brake caliper unit must be detached to allow clearance when the wheel is removed. Simply unscrew the two mounting bolts from one caliper to achieve this.

4   Remove the split pin and nut from the wheel spindle and then detach the spindle clamp from the base of the right-hand fork lower leg, where it is retained by two nuts. The wheel should be supported when removing the clamp, to prevent the left-hand fork leg spindle eye from becoming strained. Withdraw the wheel spindle and lower the wheel out of the forks.

5   Replace the front wheel by reversing the dismantling proce-

dure. After installing the wheel spindle fit the clamp so that it supports the wheel and tighten the spindle nut. Do not omit to fit a new split pin. The spindle clamp nuts must be tightened in a special sequence. Tighten the front nut fully so that the front faces of the clamp and fork leg are together. Now tighten the rear nut to the same torque setting as the front nut. The recommended torque wrench setting is 1.8 – 2.5 kgf m (13 – 18 lbf ft); this will leave a parallel gap between the clamp rear faces. Insert the speedometer cable, securing it by means of the screw. Rotating the wheel will help locate the speedometer inner cable end with the drive dogs at the end of the shaft.

3.2 Detach the speedometer cable by removing the single retaining screw and remove wheel spindle nut

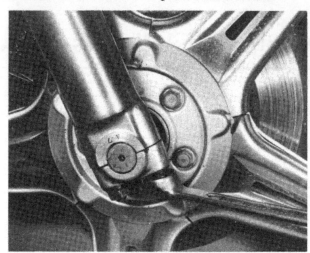

3.4 Remove the spindle clamp by releasing the two retaining nuts

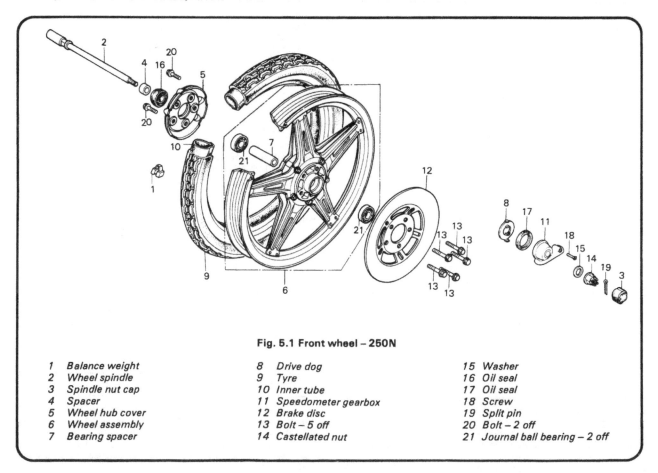

Fig. 5.1 Front wheel – 250N

1 Balance weight	8 Drive dog	15 Washer
2 Wheel spindle	9 Tyre	16 Oil seal
3 Spindle nut cap	10 Inner tube	17 Oil seal
4 Spacer	11 Speedometer gearbox	18 Screw
5 Wheel hub cover	12 Brake disc	19 Split pin
6 Wheel assembly	13 Bolt – 5 off	20 Bolt – 2 off
7 Bearing spacer	14 Castellated nut	21 Journal ball bearing – 2 off

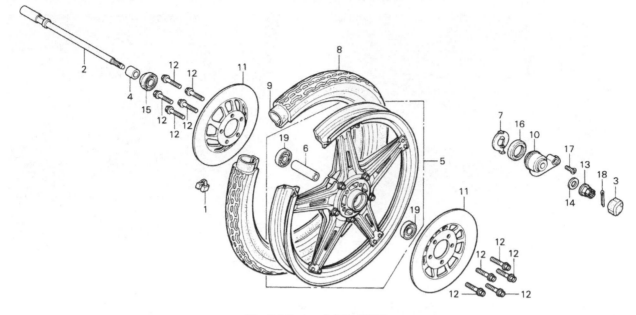

Fig. 5.2 Front wheel – 400N

1	Balance weight	6	Bearing spacer
2	Wheel spindle	7	Speedometer drive dog
3	Spindle nut cap	8	Tyre
4	Spacer	9	Inner tube
5	Wheel assembly	10	Speedometer gearbox

11	Brake disc – 2 off	16	Oil seal
12	Bolt – 10 off	17	Screw
13	Castellated nut	18	Split pin
14	Washer	19	Journal ball bearing
15	Oil seal		– 2 off

4   Front wheel disc brake: examination and renovation

1   Check the front brake master cylinder, hose and caliper unit for signs of fluid leakage duplicating the checks, where necessary on CB400N models. Pay particular attention to the condition of the hoses, which should be renewed without question if there are signs of cracking, splitting or other exterior damage.

2   Check the level of hydraulic fluid by viewing through the translucent reservoir. If the fluid is below the lower level mark, with the handlebars so placed that the reservoir is vertical, brake fluid of the correct grade should be added. The correct fluid should conform to SAE JI703 (UK) or DOT 3 (USA) specifications. **Never use engine oil** or any fluid other than that recommended. Other fluids have unsatisfactory characteristics and will rapidly destroy the seals.

3   The brake pads should be inspected for wear. Each has a red groove which marks the limit of the friction material. When this limit is reached, **both** pads must be renewed, even if only one has reached the wear mark.

4   The pad wear marks may be viewed from the rear of the caliper assembly after lifting the hinged plastic inspection cover. If the pads require renewal they may be removed from the caliper without removing the wheel, as follows:

CB250N model
5   Remove the single screw which holds the fluted cover to the caliper. Lift off the cover to expose the pads. Prise out the pad locating pin spring clip. Before withdrawing the pins, push the caliper hard over to the right. This will cause the piston to retreat into the cylinder, allowing room for the subsequent fitting of new pads.

6   After removing the two pins, the pads may be lifted from position individually. Note the shim placed against the rear face of the outer pad. Refit the new pads and the shim by reversing the dismantling procedure. Ensure that after fitting the pad pins, the retaining spring is installed so that each arm enters the drilled radial hole in the appropriate pin.

CB400N model
7   Unscrew the two bolts which pass into the caliper housing, securing it to the support bracket. The housing can now be lifted off, leaving the two pads in position in the support bracket. Displace each pad and the anti-chatter shim fitted against the rear face of the pad on the piston side of the caliper.

8   Place two new pads in position and refit the shim so the arrow mark on the shim is facing in the normal direction of disc rotation. Before replacing the caliper housing push in the piston so that sufficient clearance is given for the new pads. Pushing in the piston will cause the fluid level in the master cylinder reservoir to rise; check that it does not overflow.

9   On CB400N models the procedure should be repeated for the second caliper unit. In practice it will probably be found that both sets of pads wear at a similar rate.

4.2 Brake fluid level should be between upper and lower marks

4.4 Pad wear may be checked through inspection window

4.5 With the cover removed the pads are exposed

4.6a Note positioning of anti-chatter spring (arrowed)

4.6b The pad with the spring shim removed

4.7 Remove the two bolts (arrowed) which hold the caliper to the bracket

5 Front disc brake: overhauling the caliper unit

1   Caliper removal may be accomplished without removing the front wheel; indeed on CB400N models at least one caliper must be detached before wheel removal can be accomplished. Before commencing caliper removal the hydraulic fluid should be drained from the brake system and reservoir. Bear in mind the solvent characteristics of the fluid when doing this. Disconnect the brake hose at the banjo union on the caliper by removing the banjo bolt and allow the fluid to drain into a suitable container. Pump the handlebar lever to help expel all the fluid. On CB400N models disconnect the brake hose at the second caliper and repeat the operation.

CB250N
2   Remove the brake pads as described in the preceding Section. Unscrew the two bolts which pass through the caliper unit, and upon which the unit slides during operation. The two halves of the unit can now be separated. Push out the bolts to free the inner caliper half from the support bracket. The support bracket is secured to the fork leg by two bolts.

3   Remove the piston boot from the caliper cylinder (outer caliper half). To displace the piston, apply a blast of compressed air to the fluid inlet. Take care to catch the piston as it emerges – if dropped or prised out with a screwdriver the piston may be irreparably damaged.

4   The parts removed should be cleaned thoroughly, using only brake fluid as the liquid. Petrol, oil or paraffin will cause the various seals to swell and degrade, and should not be used under any circumstances. When the various parts have been cleaned, they should be stored in polythene bags until reassembly, so that they are kept dust free.

5   Examine the piston for score marks or other imperfections. If it has any imperfections it must be renewed, otherwise air or hydraulic fluid leakage will occur, which will impair braking efficiency. With regard to the various seals, it is advisable to renew them all, irrespective of their appearance. It is a small price to pay against the risk of a sudden and complete front brake failure. Check the slider bolts for wear, together with the holes in the support bracket in which they slide. Wear at these points will cause brake judder and poor brake release.

6   Reassemble under clinically clean conditions, by reversing the dismantling procedure. Apply a small quantity of graphite grease to the slider bolts before fitting the boots. Reconnect the hydraulic fluid pipe and make sure the union has been tightened fully. Before the brake can be used, the whole system must be refilled with brake fluid and bled of air, by following the procedure described in Section 8 of this Chapter.

CB400N

7   The dismantling and renovation procedures for the 400N brake caliper is in most respects similar to those given for the 250N, but as can be seen by comparing the illustrations of both type of caliper there are some major differences in construction. The caliper housing does not separate into two halves and the manner in which the caliper housing is mounted on the support bracket is different.

6   Removing and replacing the brake disc

1   It is unlikely that the disc will require attention until a considerable mileage has been covered, unless premature scoring of the disc has taken place thereby reducing braking efficiency. To remove the disc, first detach the front wheel as described in Section 3 of this Chapter.

2   The single disc of the CB250N is secured by five bolts which pass into the left-hand side of the hub. Similarly, each of the discs fitted to the CB400N are secured by five bolts. In neither case are locking washers or spring washers fitted into the hub.

3   The brake disc can be checked for wear and for warpage whilst the front wheel is still in the machine. Using a micrometer, measure the thickness of the disc at the point of greatest wear. If the measurement is much less than the recommended service limit of 4.0 mm (0.158 in) the disc should be renewed. Check the warpage of the disc by setting up a suitable pointer close to the outer periphery of the disc and spinning the front wheel slowly. If the total warpage is more than 0.3 mm (0.012 in) the disc should be renewed. A warped disc, apart from reducing the braking efficiency, is likely to cause juddering during braking and will also cause the brake to bind when it is not in use.

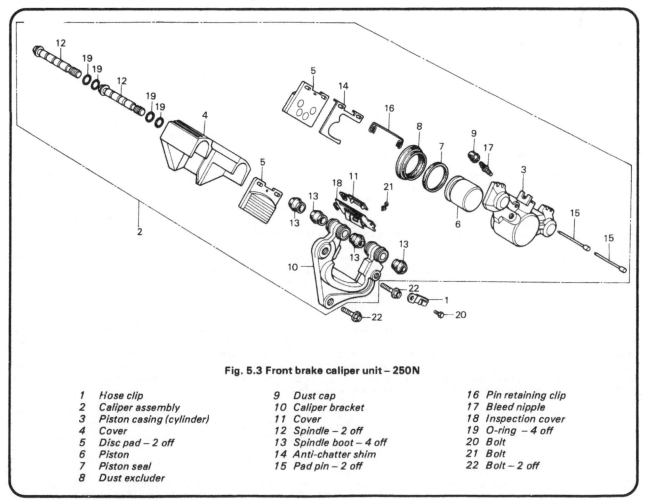

Fig. 5.3 Front brake caliper unit – 250N

1	Hose clip	9	Dust cap	16	Pin retaining clip
2	Caliper assembly	10	Caliper bracket	17	Bleed nipple
3	Piston casing (cylinder)	11	Cover	18	Inspection cover
4	Cover	12	Spindle – 2 off	19	O-ring – 4 off
5	Disc pad – 2 off	13	Spindle boot – 4 off	20	Bolt
6	Piston	14	Anti-chatter shim	21	Bolt
7	Piston seal	15	Pad pin – 2 off	22	Bolt – 2 off
8	Dust excluder				

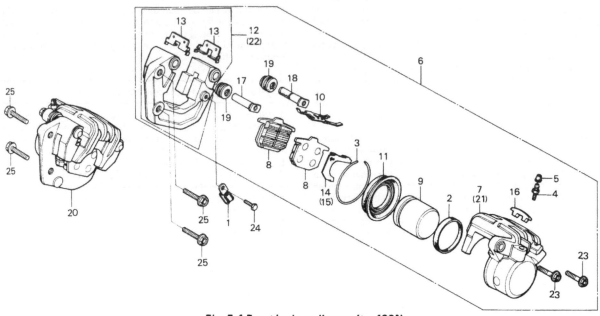

Fig. 5.4 Front brake caliper unit – 400N

1 Hose clip	8 Disc pad – 2 off	15 Anti-chatter shim	21 Cover and piston casing
2 Piston seal	9 Piston	16 Inspection cover	(cylinder)
3 Dust cover clip	10 Pad spring	17 Spindle	22 Caliper bracket assembly
4 Bleed nipple	11 Dust excluder	18 Spindle	23 Bolt – 2 off
5 Dust cap	12 Caliper bracket assembly	19 Spindle boot – 2 off	24 Bolt
6 Caliper assembly	13 Pad retaining clip – 2 off	20 Caliper and bracket	25 Bolt – 4 off
7 Cover and piston casing	14 Anti-chatter shim	assembly complete	
(cylinder)			

7 Master cylinder: examination and renovation

1 The master cylinder is unlikely to give trouble unless the machine has been stored for a long period or until a considerable mileage has been covered. The usual signs of trouble are leakage of fluid, causing a gradual fall in the fluid level and bad braking performance.

2 To gain access to the master cylinder, commence the dismantling operation by attaching a bleed tube to the caliper unit bleed valve. Open the bleed valve one complete turn, then operate the front brake lever until all the hydraulic fluid is pumped out of the reservoir. Close the bleed valve, remove the bleed tube and store the fluid in a closed container for subsequent reuse.

3 Remove the banjo bolt and fluid hose from the end of the master cylinder unit and remove the lever pivot bolt, the lever and the front brake switch. Detach the master cylinder from the handlebars by removing the two bolts and the securing clamp.

4 Access is now available to the piston and the cylinder and it is possible to remove the piston assembly, together with the relevant seals. Remove the circlip and sealing boot, followed by the next internal clip. The remainder of the components can be pushed out. Take note of the way in which the seals are fitted because they must be replaced in the same order and position. Failure to observe this necessity will result in brake failure.

5 Clean the master cylinder and piston with either hydraulic fluid or alcohol. On no account use abrasives or other solvents such as petrol. If any signs of damage or wear are evident, renewal is necessary. It is not practicable to reclaim either the piston or the cylinder bore.

6 Soak the new seals in hydraulic fluid for about 15 minutes prior to fitting, then reassemble the parts in exactly the same order, using the reversal of the dismantling procedure. Lubricate with hydraulic fluid and make sure the feathered edges of the seals are not damaged.

7 Refit the assembled master cylinder to the handlebars and reconnect the handlebar lever and hose. Refill the reservoir with hydraulic fluid and bleed the entire system by following the procedure described in Section 9 of this Chapter.

8 Check that the brake is working correctly before taking the machine on the road, to restore pressure and align the pads correctly. Use the brake gently for the first 50 miles or so to let the new components bed down correctly.

9 It should be emphasised that repairs to the master cylinder are best entrusted to a Honda Service Agent, or alternatively, that the defective parts should be replaced by a new unit. Dismantling and reassembly requires a certain amount of skill and it is imperative that the entire operation is carried out under surgically clean conditions.

8 Front disc brake: bleeding the hydraulic system

1 Removal of all the air from the hydraulic system is essential for the efficiency of the braking system. Air can enter the system due to leaks or when any part of the system has been dismantled for repair or overhaul. Topping the system up will not suffice, as air pockets will still remain, even small amounts causing dramatic loss of brake pressure.

2 Check the level in the reservoir, and fill almost to the top. Again, beware of spilling the fluid on to painted or plastic surfaces.

3 Place a clean jar below the brake caliper unit and attach a clear plastic tube from the caliper bleed screw to the container. Place some clean hydraulic fluid in the container so that the pipe is always immersed below the surface of the fluid.

4 Unscrew the bleed screw one complete turn and pump the handlebars lever slowly. As the fluid is ejected from the bleed screw the level in the reservoir will fall. Take care that the level does not drop too low whilst the operation continues, otherwise air will re-enter the system, necessitating a fresh start.

5   Continue the pumping action with the lever until no further air bubbles emerge from the end of the plastic pipe. Hold the brake lever against the handlebars and tighten the caliper bleed screw. Remove the plastic tube after the bleed screw is closed.

6   Check the brake action for sponginess, which usually denotes there is still air in the system. If the action is spongy, continue the bleeding operation in the same manner, until all traces of air are removed. On CB400N models the procedure should be repeated on the second caliper.

7   When all traces of air have been removed from the system, top up the reservoir and refit the diaphragm and cap or cover, as appropriate. Check the entire system for leaks, and check also that the brake system in general is functioning efficiently before using the machine on the road.

8   Brake fluid drained from the system will almost certainly be contaminated, either by foreign matter or more commonly by the absorption of water from the air. All hydraulic fluids are to some degree hygroscopic, that is, they are capable of drawing water from the atmosphere, and thereby degrading their specifications. In view of this, and the relative cheapness of the fluid, old fluid should always be discarded.

8.3 Attach a clear plastic tube to the caliper bleed screw

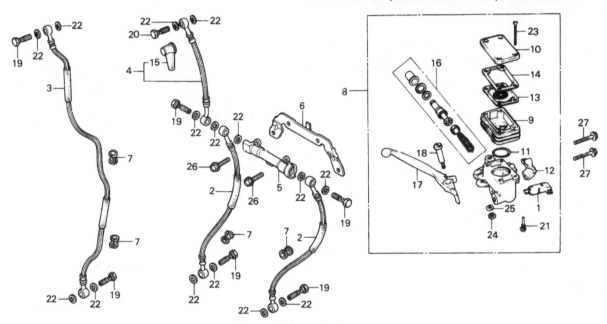

Fig. 5.5 Master cylinder

1 Stop lamp switch	6 Two-way joint stay (400N)	14 Diaphragm plate
2 Brake hose assembly – 2 off (400N)	7 Grommet – 2 off	15 Master cylinder boot
	8 Master cylinder assembly	16 Piston assembly
3 Brake hose assembly (250N)	9 Oil reservoir	17 Brake lever
	10 Reservoir cap	18 Pivot bolt
4 Master cylinder hose (400N)	11 O-ring	19 Bolt – 4 off (2 off, 250N)
	12 Cylinder clamp	20 Bolt (not 250N)
5 Two-way joint (400N)	13 Diaphragm	21 Screw

22 Washer – 11 off (4 off, 250N)	
23 Screw – 4 off	
24 Nut	
25 Washer	
26 Bolt – 2 off (not 250N)	
27 Bolt – 2 off	

9 Front wheel bearings: examination and replacement

1   Place the machine on the centre stand and remove the front wheel as described in Section 4 of this Chapter. Remove the speedometer gearbox, drive gear and oil seal on the right-hand side of the hub and remove the spacer and oil seal from the left-hand side.

2   The wheel bearings can now be tapped out from each side with the use of a suitable long drift. Careful and even tapping will prevent the bearing 'tying' and damage to the races.

3   Remove all the old grease from the hub and bearings, giving the latter a final wash in petrol. When the bearings are clean lubricate them sparingly with a very light oil. Check the bearings

for play and roughness when they are spun by hand. All used bearings will emit a small amount of noise when spun but they should not chatter or sound rough. If there is any doubt about the condition of the bearings they should be renewed.

4   Before replacing the bearings pack them with high melting point grease. Do not overfill the hub centre with grease as it will expand when hot and may find its way past the oil seals. The hub space should be about $\frac{2}{3}$ full of grease. Drift the bearings in, using a soft drive on the outside ring of the bearing. Do not drive the centre ring of the bearing or damage will be incurred. Replace the oil seals carefully, drifting them into place with a thick walled tube of approximately the same dimension as the oil seal. A large socket spanner is ideal.

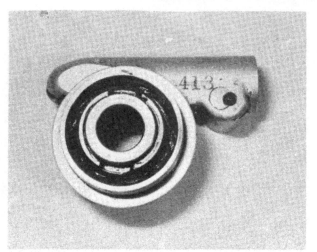

9.1a The speedometer gearbox requires occasional lubrication

9.1b The oil seal and speedometer gearbox drive dog in situ

10 Rear wheel: examination, removal and renovation

1   Place the machine on the centre stand so that the rear wheel is raised clear of the ground. Check for rim alignment and damage to the rim and loose or broken spokes by following the procedure relating to the front wheel, as described in Section 4 of this Chapter.

11 Rear wheel: removal and replacement

1   Place the machine on the centre stand so that the rear wheel is raised clear of the ground. Disconnect the rear chain at the master link by removing the spring retainer; the chain need not be run off the gear box sprocket.
2   Unscrew the adjuster nut from the end of the brake operating rod and then depress the brake pedal so that the rod leaves the trunnion in the operating arm. Push out the trunnion and fit it, together with the adjuster nut, to the rod, to avoid loss. Detach the brake torque arm from the back plate by removing the securing split pin and unscrewing the nut.
3   Remove the split pin from the wheel spindle and unscrew the castellated wheel spindle nut. Support the weight of the wheel and push out the spindle. Knock the wheel spacer out from between the brake back plate and the fork end. The wheel can now be lifted clear from between the fork ends.
4   Replace the wheel by reversing the dismantling procedure. The chain adjusters must not be omitted when inserting the wheel spindle. Before tightening the spindle nut fit the chain and check that the wheel is aligned correctly. See the instructions for chain adjustment in Section 18. Do not forget to secure the brake torque arm bolt by means of the split pin. If the torque arm becomes detached in service, the brake will lock on first application, causing an accident.

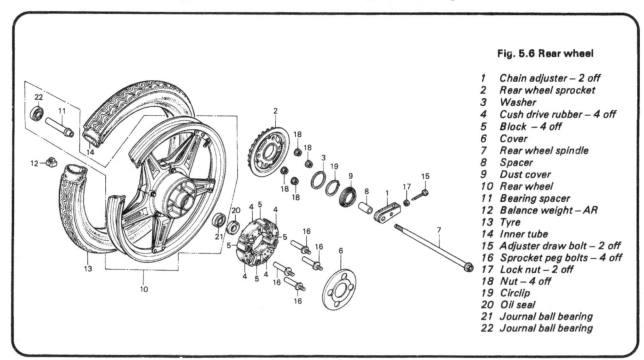

Fig. 5.6 Rear wheel

1   Chain adjuster – 2 off
2   Rear wheel sprocket
3   Washer
4   Cush drive rubber – 4 off
5   Block – 4 off
6   Cover
7   Rear wheel spindle
8   Spacer
9   Dust cover
10   Rear wheel
11   Bearing spacer
12   Balance weight – AR
13   Tyre
14   Inner tube
15   Adjuster draw bolt – 2 off
16   Sprocket peg bolts – 4 off
17   Lock nut – 2 off
18   Nut – 4 off
19   Circlip
20   Oil seal
21   Journal ball bearing
22   Journal ball bearing

11.2a Remove the brake adjuster nut and ...

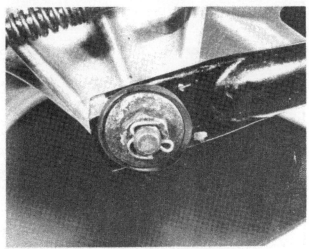

11.2b ... the brake torque arm secured by a split pin

11.3a Remove split pin and unscrew the castellated nut

11.3b Slide spindle through chain adjuster

12 Rear brake: examination and renovation

1   The brake back plate complete with the brake shoes may be withdrawn from the wheel hub after the wheel has been removed as described in Section 11.
2   Examine the condition of the brake linings. If they are wearing thin or unevenly, the brake shoes should be renewed. The linings are bonded to the brake shoes and cannot be supplied separately.
3   To remove the brake shoes, turn the brake operating lever so that the brake is in the full-on position. Pull the brake shoes apart to free them from their operating cam and from the pivot on which the fixed ends bear. Then pull them upward in a V formation so that they can be lifted away together with the return springs. When they are well clear of the brake plate, the return springs can be detached.
4   Check the inner surface of the brake drum, on which the brake shoes bear. The surface should be free from score marks or indentations, otherwise reduced braking efficiency will be inevitable. Remove all traces of brake lining dust and wipe with a rag soaked in petrol, to remove all traces of grease and oil.
5   Before replacing the brake shoes, check that the brake operating cam is working smoothly and not binding at its pivot. The cam can be removed for greasing by detaching the operating arm from the splined shaft end. The operating arm has splines which engage with similar splines on the cam shaft end;

mark the operating arm and the splined shaft to aid correct relocation.
6   To reassemble the brake shoes on the brake plate, fit the return springs and pull the shoes apart whilst holding them in the form of a V, facing upwards. If they are now located with the brake operating cam and fixed pivot, they can be pushed back into position by pressing downward. Do not use force, or there is risk of distorting the shoes.

13 Adjusting the rear brake

All models
1   Adjustment of the rear brake is correct when there is 20 – 30 mm ($\frac{3}{4}$ – $1\frac{1}{4}$ in) up and down movement measured at the rear brake pedal foot piece, between the fully 'off' and 'on' position.
2   If, when the brake is fully applied, the angle between the brake arm and the operating rod is more than 90° the brake arm should be pulled off the camshaft, after loosening the pinch bolt. Reset the brake arm so that the right angle is produced.
3   The height of the brake pedal may be adjusted to suit the preference of the rider by means of a bolt and locknut fitted to a lug to the rear of the pedal. Note that it may be necessary to adjust the height-setting of the stop lamp switch after adjustment of the brake pedal position.

98

Fig. 5.7 Rear brake

1 Chain adjuster – 2 off
2 Wheel spacer
3 Brake backplate assembly
4 Brake shoe – 2 off
5 Brake cam
6 Brake shoe spring – 2 off
7 Brake operating arm
8 Rubber arm stopper
9 Dust seal
10 Wear indicator
11 Chain adjusting bolt
12 Bolt
13 Castellated nut
14 Nut
15 Washer
16 Nut
17 Nut
18 Split pin
19 Split pin
20 Split pin
21 Bolt

12.1a The brake plate may be pulled out of the drum

12.1b Pull the brake shoes away in a 'V' form

12.1c Check the drum surface for scoring. Remove dust

13.3a Brake pedal height may be adjusted by bolt and locknut

13.3b Adjustment of the stop lamp switch may be necessary

14 Rear wheel bearings: removal and replacement

1   To gain access to the rear wheel bearings, the wheel must be removed from the machine and the brake back plate withdrawn from the brake drum.

2   Knock out the bearings, using the same procedure as described for the front wheel, starting with the right-hand bearing and then the left-hand bearing. The remarks on bearing testing and regreasing described for the front wheel apply similarly to the rear wheel.

15 Rear wheel sprocket: examination and replacement

1   The rear wheel sprocket is retained on the left-hand side of the hub by a large circlip, and is located by four pegs which pass into the cush drive in the hub and are retained by hexagonal nuts.

2   To remove the sprocket, detach the large circlip and the spacing plate which lies below. The circlip is protected by a rubber cover which should be prised from position. Pull the sprocket off the hub, together with the four drive pins.

3   Check the condition of the sprocket teeth. If they are hooked, chipped or badly worn, the sprocket must be renewed. It is considered bad practice to renew one sprocket on its own. The final drive sprockets should always be renewed as a pair and a new chain fitted, otherwise rapid wear will necessitate even earlier renewal on the next occasion.

4   The sprocket may be refitted by reversing the dismantling procedure. It is important that the recesses in the rear of the sprocket are engaged correctly by the milled flats on each cush drive pin.

16 Rear wheel cush drive: examination and renovation

1   All models are fitted with a cush drive arrangement in the rear hub which allows a small amount of movement between the hub and the drive sprocket to help damp out transmission shock loads.

2   The pins at the rear of the sprocket pass into cast aluminium blocks which are located in a moulded rubber buffer. After removal of the circlip and sprocket, the cush drive cover may be lifted away to gain access to the blocks and buffers. If the buffers have become compacted they should be renewed.

14.2a The wheel bearings driven out

14.2b Wheel spacer and bearing ...

14.2c ... fit into hub centre

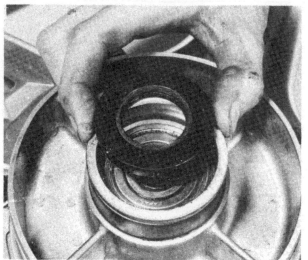

14.2d Do not omit to replace oil seal

15.2a Prise off the rubber cover to allow ...

15.2b ... removal of the sprocket retaining circlip

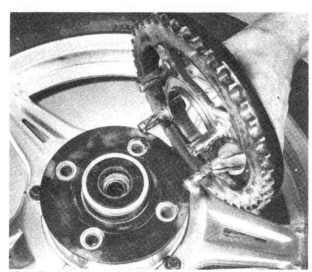

15.2c The sprocket may be lifted away, complete with pins

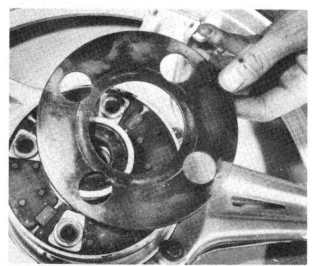

16.2a Removal of the cush drive cover enables access to blocks and buffers

16.2b The pin blocks are mounted as shown and ...

16.2c ... should be fitted as shown

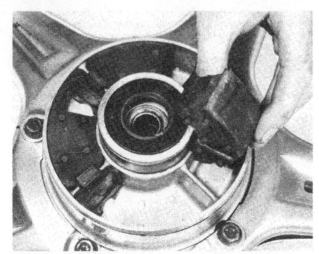

16.2d Check the rubber blocks for compaction or perishing

17 Chain: examination, lubrication and adjustment

1   The final drive chain is fully exposed apart from the protection given by a short chainguard along the upper run, and if not properly maintained will have a short life. A worn chain will cause rapid wear of the sprockets and they too will need replacement.

2   The chain tension will need adjustment at regular intervals, to compensate for wear. This is accomplished by loosening the rear wheel nut, which is secured by a split pin, and loosening the large nut holding the cush drive/sprocket hub hollow spindle, with the machine on the centre stand. The brake torque arm nuts should be loosened, but it is not necessary to remove the spring security pins.

3   Slacken the locknuts on the chain adjusters on the fork ends. Screw the adjusters inwards an equal amount to tighten the chain. The tension is correct if there is 15 – 20 mm (⅝ in – ¾ in) up and down movement in the centre of the lower chain run. Always check the chain when it is at its tightest point; a chain rarely wears evenly. This may be accomplished by turning the wheel whilst applying a finger to the lower chain run. The tightest point is easily found.

4   Always adjust the draw bolts an even amount so that correct wheel alignment is preserved. The fork ends are marked with a series of vertical lines to provide a visual check. If desired, wheel alignment can be checked by running a plank of wood parallel to the machine so that it touches both walls of the rear tyre. If wheel alignment is correct, it should be equidistant from either side of the front wheel tyre when tested on both sides of the rear wheel; it will not touch the front tyre because this tyre has a smaller cross section. See the accompanying diagram.

5   Do not run the chain overtight to compensate for uneven wear. A tight chain will place excessive stresses on the gearbox and rear wheel bearings leading to their early failure. It will also absorb a surprising amount of power.

6   After a period of running the chain will require lubrication. Lack of oil will accelerate the rate of wear of both chain and sprockets and will lead to harsh transmission. The application of engine oil will act as a temporary expedient, but it is preferable to remove the chain and immerse it in a molten lubricant such as Linklyfe or Chainguard after it has been cleaned in a paraffin bath. These latter lubricants achieve better penetration of the chain links and rollers and are less likely to be thrown off when the chain is in motion.

7   To check whether the chain is due for replacement, lay it lengthwise in a straight line and compress it endwise until all play is taken up. Anchor one end, then pull in the opposite direction to take up the play which develops. If the chain extends by more that ¼ inch per foot, it should be renewed in conjunction with the sprockets. Note that this check should ALWAYS be made after the chain has been washed out, but before any lubricant is applied, otherwise the lubricant may take up some of the play.

8   When fitting the chain on the machine, make sure the spring link is positioned correctly with the closed end facing the direction of travel.

9   Replacement chains are now available in standard metric sizes from Renold Limited, the British chain manufacturer. When ordering a new chain, always quote the size, the number of chain links and the type of machine to which the chain is to be fitted.

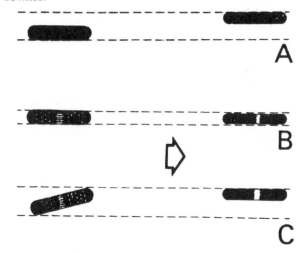

Fig. 5.8 Method of checking wheel alignment

A & C Incorrect;   B Correct

18 Tyres: removal and replacement

1   At some time or other the need will arise to remove and replace the tyres, either as the result of a puncture or because a

renewal is required to offset wear. To the inexperienced, tyre changing represents a formidable task yet if a few simple rules are observed and the technique learned, the whole operation is surprisingly simple.

2   To remove the tyre from either wheel, first detach the wheel from the machine by following the procedure in Section 4 or 12 of this Chapter, depending on whether the front or the rear wheel is involved. Deflate the tyre by removing the valve insert and when it is fully deflated, push the bead of the tyre away from the wheel rim on both sides so that the bead enters the centre well of the rim. Remove the locking cap and push the tyre valve into the tyre itself.

3   Insert a tyre lever close to the valve and lever the edge of the tyre over the outside of the wheel rim. Very little force should be necessary; if resistance is encountered it is probably due to the fact that the tyre beads have not entered the well of the wheel rim all the way round the tyre.

4   Once the tyre has been edged over the wheel rim, it is easy to work around the wheel rim so that the tyre is completely free on one side. At this stage, the inner tube can be removed.

5   Working from the other side of the wheel, ease the other edge of the tyre over the outside of the wheel rim which is furthest away. Continue to work around the rim until the tyre is freed completely from the rim.

6   If a puncture has necessitated the removal of the tyre, re-inflate the inner tube and immerse it in a bowl of water to trace the source of the leak. Mark its position and deflate the tube. Dry the tube and clean the area around the puncture with a petrol soaked rag. When the surface has dried, apply the rubber solution and allow this to dry before removing the backing from the patch and applying the patch to the surface.

7   It is best to use a patch of the self-vulcanising type which will form a very permanent repair. Note that it may be necessary to remove a protective covering from the top surface of the patch, after it has sealed in position. Inner tubes made from synthetic rubber may require a special type of patch and adhesive if a satisfactory bond is to be achieved.

8   Before refitting the tyre, check the inside to make sure that the agent which caused the puncture is not trapped. Check the outside of the tyre, particularly the tread area, to make sure nothing is trapped that may cause a further puncture.

9   If the inner tube has been patched on a number of past occasions, or if there is a tear or large hole, it is preferable to discard it and fit a new one. Sudden deflation may cause an accident, particularly if it occurs with the front wheel.

10   To replace the tyre, inflate the inner tube sufficently for it to assume a circular shape but only just. Then push it into the tyre so that it is enclosed completely. Lay the tyre on the wheel at an angle and insert the valve through the rim tape and the hole in the wheel rim. Attach the locking cap on the first few threads, sufficient to hold the valve captive in its correct location.

11   Starting at the point furthest from the valve, push the tyre bead over the edge of the wheel rim until it is located in the central well. Continue to work around the tyre in this fashion until the whole of one side of the tyre is on the rim. It may be necessary to use a tyre lever during the final stages.

12   Make sure that there is no pull on the tyre valve and again commencing with the area furthest from the valve, ease the other bead of the tyre over the edge of the rim. Finish with the area close to the valve, pushing the valve up into the tyre until the locking cap touches the rim. This will ensure the inner tube is not trapped when the last section of the bead is edged over the rim with a tyre lever.

13   Check that the inner tube is not trapped at any point. Re-inflate the inner tube and check that the tyre is seating correctly around the wheel rim. These should be a thin rib moulded around the wall of the tyre on both sides which should be equidistant from the wheel rim at all points. If the tyre is unevenly located on the rim, try bouncing the wheel when the tyre is at the recommended pressure. It is probable that one of the beads has not pulled clear of the centre well.

14   Always run the tyres at the recommended pressures and never under or over-inflate. The correct pressures for solo use are given in the Specifications Section of this Chapter. If a pillion passenger is carried, increase the rear tyre pressure only by approximately 4 psi.

15   Tyre replacement is aided by dusting the side walls, particularly in the vicinity of the beads, with a liberal coating of French chalk. Washing up liquid can also be used to good effect, but this has the disadvantage of causing the inner surfaces of the wheel rim to rust.

16   Never fit a tyre which has a damaged tread or side walls. Apart from the legal aspects, there is a very great risk of a blow-out, which can have serious consequences.

19 Fault diagnosis: wheels, brakes and tyres; on page 104

Tyre changing sequence - tubed tyres

  A Deflate tyre. After pushing tyre beads away from rim flanges push tyre bead into well of rim at point opposite valve. Insert tyre lever adjacent to valve and work bead over edge of rim.

B Use two levers to work bead over edge of rim. Note use of rim protectors

 C Remove inner tube from tyre

 D When first bead is clear, remove tyre as shown

 E When fitting, partially inflate inner tube and insert in tyre

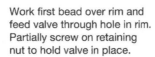 **F** Work first bead over rim and feed valve through hole in rim. Partially screw on retaining nut to hold valve in place.

 G Check that inner tube is positioned correctly and work second bead over rim using tyre levers. Start at a point opposite valve.

H Work final area of bead over rim whilst pushing valve inwards to ensure that inner tube is not trapped

19 Fault diagnosis: Wheels, brakes and tyres

Symptom	Cause	Remedy
Handlebars oscillate at low speeds	Buckle or flat in wheel rim, most probably front wheel	Check rim alignment by spinning wheel. Renew complete wheel (Comstar wheel).
	Tyre not straight on rim	Check tyre alignment.
Machine lacks power and accelerates poorly	Rear brake binding	Warm brake drum provides best evidence. Re-adjust brake.
Rear brake grabs when applied gently	Ends of brake shoes not chamfered	Chamfer with file.
	Elliptical brake drum	Lightly skim in lathe (specialist attention required).
Front brake feels spongy	Air in hydraulic system	Bleed brake.
Brake pull-off sluggish	Rear brake; broken or worn shoe return springs	Remove back plate and renew the springs.
	Front brake; sticking pistons in brake caliper	Overhaul caliper unit.
Harsh transmission	Worn or badly adjusted final drive chain	Adjust or renew as necessary.
	Hooked or badly worn sprockets	Renew as a pair.
	Worn or deteriorating cush drive rubbers	Renew rubbers.

Chapter 6 Electrical system

Contents

Specifications

Battery
Make	Yuasa
Capacity	12 volt, 12 Ah
Earth	Negative

Alternator
Make	Nippon Denso (250 model), Hitachi (400 model)
Type	Rotating permanent magnet rotor, multi-coil stator incorporating CDI ignition power source and timer
Output	150 watts @ 5000 rpm (400N) 130 watts @ 5000 rpm (250N)

Starter motor
Brush length	11·0 – 12·5 mm (0·43– 0·49 in)
Service limit	5·5 mm (0·21 in)

Bulbs
Headlamp	45/40W (250N) 60/55W (400N)
Tail/brake lamp	5/21W
Instrument lights	3·4W
Pilot light	4W
High beam indicator	3·4W
Neutral indicator	3·4W
Oil pressure warning	3·4W
Flashing indicators	21W
Flasher indicator bulb	3·4W

All bulbs rated 12 volt

1 General description

The models covered by this manual are fitted with a 12 volt electrical system. The circuit comprises a crankshaft mounted, permanent magnet alternator and a combined solid-state regulator rectifier unit. The regulator maintains the output to within a specified limit to prevent overcharging and the rectifier converts the ac (alternating current) output to dc (direct current) to enable the lights and ancillary equipment to be powered and to allow the battery to be charged. The alternator consists of a multi-coil stator bolted to the left-hand casing and a permanent magnet rotor. Included in the alternator are the various components which relate to the CDI ignition system.

2 Crankshaft alternator: checking the output

1 The output from the alternator mounted on the end of the crankshaft can be checked only with specialised test equipment of the multimeter type. It is unlikely that the average

owner/rider will have access to this equipment or instruction in its use. In consequence, if the performance of the alternator is in any way suspect, it should be checked by a Honda Service Agent or an auto-electrical specialist.

2 If a multimeter is available, a general check on the alternator may be carried out as follows. Connect a dc voltmeter across the two battery terminals, and install an ammeter in the battery positive lead. Disconnect the black lead from the regulator/rectifier unit. Start the engine and check the readings on both meters, which should be as follows:

Charging starts	Output at 5,000 rpm
CB400 1200 rpm max	10A min/14.5V
CB250 1500 rpm max	8A min/14.5V

Repeat the check with the headlight main beam switched on.

Charging starts	Output at 5000 rpm
CB400 1200 rpm max	5A min/14.5V
CB250 2300 rpm max	3A min/14.5V

3   Note that this check must be made with a fully charged battery and after the engine has been allowed to reach normal working temperature.

4   If the voltage and current is correct, it may be assumed that the system is functioning properly. A marked reduction in output may be a result of damaged windings in the stator coils or broken leads. These may be checked for continuity and resistance without removing the alternator from the machine, as follows.

5   Remove the left-hand side cover from the frame so that access to the rectifer/regulator unit is made. Disconnect the yellow wires from the alternator at their three-pin block connector.

6   Using a multimeter set to the resistance function, check for continuity between each yellow wire and the other two in turn, then check that there is no continuity between each yellow wire and a good earth point on the crankcase. If any wire is isolated from the others, or if any is shorted to earth the stator is faulty and must be renewed. Check first however that the fault is not due to a damaged or a broken wire; in some circumstances this can be repaired. Due to the high replacement cost of an alternator, particularly where a CDI generator coil is incorporated, it is worth having the unit checked by a competent auto-electrician before consigning the assembly to the scrap bin.

7   If the resistance tests are satisfactory, the regulator/rectifer unit should be examined as described in the following Section.

---

3   Regulator/rectifier: testing

1   This component is a heavily-finned sealed metal unit bolted to a bracket behind the left-hand side panel. If the unit is found to be damaged or faulty, it must be renewed, even if only one side is affected; repairs are not possible.

2   To check the regulator side, a voltmeter must be connected across the battery terminals. With the engine running, the voltage across the battery should be between 14 – 15V.

3   To test the rectifier, separate the leads from the unit at the two block connectors. Using a multimeter set to the appropriate resistance scale, check for continuity between the green wire and each yellow wire and between the red/white wire and each yellow wire, then reverse the meter probes and check for continuity in the opposite direction. In the normal direction of current flow the meter should show very little resistance (approx 5 – 40 ohm) but in the reverse direction much greater resistance (approx 2 K ohms) should be measured. If any of the twelve tests does not produce the expected result then that particular diode is faulty and the complete unit must be renewed.

3.1 Finned regulator/rectifier unit and main block connectors

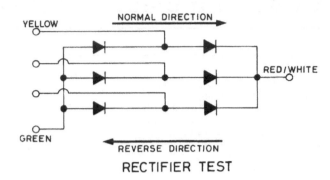

RECTIFIER TEST

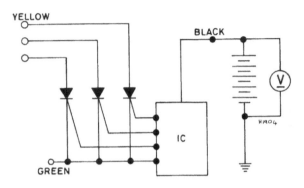

REGULATOR TEST

Fig. 6.1 Regulator/rectifier test

---

4   Battery: examination and maintenance

1   Both models are fitted with a lead-acid battery of 12Ah capacity.

2   The transparent plastic case of the battery permits the upper and lower levels of the electrolyte to be observed when the battery is pulled from its housing below the dualseat. Access to the battery may be gained by detaching the frame right-hand side cover. Maintenance is normally limited to keeping the electrolyte level between the prescribed upper and lower limits and by making sure the vent pipe is not blocked. The lead plates and their separators can be seen through the transparent case, a further guide to the general condition of the battery.

3   Unless acid is spilt, as may occur if the machine falls over, the electrolye should always be topped up with distilled water, to restore the correct level. If acid is spilt on any of the machine, it should be neutralised with an alkali such as washing soda and washed away with plenty of water, otherwise serious corrosion will occur. Top up with sulphuric acid of the correct specific gravity (1.260 – 1.280) only when spillage has occurred. Check that the vent pipe is well clear of the frame tubes or any of the other cycle parts, for obvious reasons.

4.1 Battery is easily accessible after removal of right-hand frame cover

5 Battery: charging procedure

1   The normal charging rate for any battery is $\frac{1}{10}$ the rated capacity. Hence the charging rate for the 12 Ah battery is 1.2 amps. A slightly higher rate of charge may be used in an emergency. The higher charge rate should, if possible, be avoided since it will shorten the working life of the battery.

2   Make sure that the battery charger connections are correct, red to positive and black to negative. It is preferable to remove the battery from the machine whilst it is being charged and to remove the vent plug from each cell. When the battery is reconnected to the machine, the black lead must be connected to the negative terminal and the red lead to positive. This is most important, as the machine has a negative earth system. If the terminals are inadvertently reversed, the electrical system will be damaged permanently.

6 Fuse: location and replacement

1   A bank of fuses is contained within a small plastic box located near the regulator. Three fuses are used, the main one being of 15A rating and the remaining two of 7A rating.

2   Before replacing a fuse that has blown, check that no obvious short circuit has occurred, otherwise the replacement fuse will blow immediately it is inserted. It is always wise to check the electrical circuit thoroughly, to trace the fault and eliminate it.

3   When a fuse blows while the machine is running and no spare is available, a 'get you home' remedy is to remove the blown fuse and wrap it in silver paper before replacing it in the fuse holder. The silver paper will restore the electrical continuity by bridging the broken fuse wire. This expedient should NEVER be used if there is evidence of short circuit or other major electrical fault, otherwise more serious damage will be caused. Replace the 'doctored' fuse at the earliest possible opportunity, to restore full circuit protection.

7 Starter motor: removal, examination and replacement

1   An electric starter motor, operated from a small push-button on the right-hand side of the handlebars, provides an alternative and more convenient method of starting the engine, without having to use the kickstart. The starter motor is mounted at the front of the crankcase, immediately below the front down tube. Current is supplied from the battery via a heavy duty solenoid switch and a cable capable of carrying the very high current demanded by the starter motor on the initial start-up.

2   The starter motor drives a free running clutch via an idler pinion. The clutch ensures the starter motor drive is disconnected from the pump drive immediately the engine starts. It operates on the centrifugal principle; spring loaded rollers take up the drive until the centrifugal force of the rotating engine overcomes their resistance and the drive is automatically disconnected.

3   The starter motor may be removed while the engine is in the frame. Disconnect the battery leads to isolate the electrical system and then detach the heavy cable from the terminal on the starter motor body. The protecting rubber boot will have to be prised off the terminal to gain access to the nut.

4   The starter motor is secured to the crankcase by two bolts which pass through the left-hand end of the motor casing. When these bolts are withdrawn, the motor can be prised out of position and lifted away. If necessary, a thin wooden wedge placed between the motor casing and crankcase may be used to push the starter motor backwards so that the motor boss leaves the hole in the casing.

5   The parts of the starter motor most likely to require attention are the brushes. The end cover is retained by the two long screws which pass through the lugs cast on both end pieces. If the screws are withdrawn, the end cover can be lifted away and the brush gear exposed.

6   Lift up the spring clips which bear on the end of each brush and remove the brushes from their holders. Each brush should have a length of 12.5 mm (0.5 in). The minimum allowable brush length is 5.5 mm (0.22 in). If the brush is shorter it must be renewed.

7   Before the brushes are replaced, make sure that the commutator is clean. The commutator is the ring of copper segments on which the brushes bear. Clean the commutator with a strip of glass paper. Never use emery cloth or 'wet-and-dry' as the small abrasive fragments may embed themselves in the soft copper of the commutator and cause excessive wear of the brushes. Finish off the commutator with metal polish to give a smooth surface and finally wipe the segments over with a methylated spirits soaked rag to ensure a grease free surface. Check that the mica insulators, which lie between the segments of the commutator, are undercut. The standard groove depth is 0.5 – 0.8 mm (0.02 – 0.03 in), but if the average groove depth is less than this the armature should be renewed or returned to a Honda Service Agent for re-cutting.

8   Replace the brushes in their holders and check that they slide quite freely. Make sure the brushes are replaced in their original positions because they will have worn to the profile of the commutator. Replace and tighten the end cover, then replace the starter motor and cable, and where necessary the oil pressure switch and lead, by reversing the dismantling procedure.

8 Starter solenoid switch: function and location

1   The starter motor switch is designed to work on the electro-magnetic principle. When the starter motor button is depressed, current from the battery passes through windings in the switch solenoid and generates an electro-magnetic force which causes a set of contact points to close. Immediately the points close, the starter motor is energised and a very heavy current is drawn from the battery.

2   This arrangement is used for at least two reasons. Firstly, the starter motor current is drawn only when the button is depressed and is cut off again when pressure on the button is released. This ensures minimum drainage on the battery. Secondly, if the battery is in a low state of charge, there will not be sufficient current to cause the solenoid contacts to close. In consequence, it is not possible to place an excessive drain on the battery which, in some circumstances, can cause the plates to overheat and shed their coatings. If the starter will not

operate, first suspect a discharged battery. This can be checked by trying the horn or switching on the lights. If this check shows the battery to be in good shape, suspect the starter switch which should come into action with a pronounced click. It is located behind the left-hand side panel and can be identified by the heavy duty starter cable connected to it. It is not possible to effect a satisfactory repair if the switch malfunctions; it must be renewed.

9 Headlamp: replacing the bulbs and adjusting beam height

1   In order to gain access to the headlamps bulbs it is necessary to first remove the rim, complete with the reflector and headlamp glass. The rim is retained by two screws which pass through the headlamp shell just below the two headlamp mounting bolts.

2   Pull the headlamp bulb socket from the rear bulb holder. The headlamp bulb is retained by a spring loaded collar. To release the collar, depress and then twist it in an anti-clockwise direction. The collar, spring and bulb may be lifted from position.

A reflector that accepts a pilot bulb is fitted to all models delivered to countries or states where parking lights are a statutory requirement. The pilot bulb is held in the bulb holder by a bayonet fixing.

3   Beam height on all models is effected by tilting the headlamp shell after the mounting bolts have been loosened slightly. On the CB400N the horizontal alignment of the beam can be adjusted by altering the position of the screw which passes through the headlamp rim. The screw is fitted at the 9 o'clock position when viewed from the front of the machine. Turning the screw in a clockwise direction will move the beam direction over to the left-hand side.

4   In the UK, regulations stipulate that the headlamps must be arranged so that the light will not dazzle a person standing at a distance greater than 25 feet from the lamp, whose eye level is not less than 3 feet 6 inches above that plane. It is easy to approximate this setting by placing the machine 25 feet away from a wall, on a level road, and setting the beam height so that it is concentrated at the same height as the distance of the centre of the headlamp from the ground. The rider must be seated normally during this operation and also the pillion passenger, if one is carried regularly.

9.1 Headlamp rim is secured by two screws passing through the shell

9.2a Headlamp bulb socket is a push fit

9.2b Bayonet collar locates the bulb holder; bulb is a bayonet fit

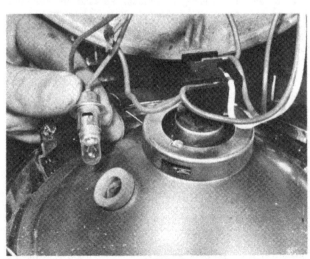

9.2c Pilot bulb holder is a push fit in the reflector

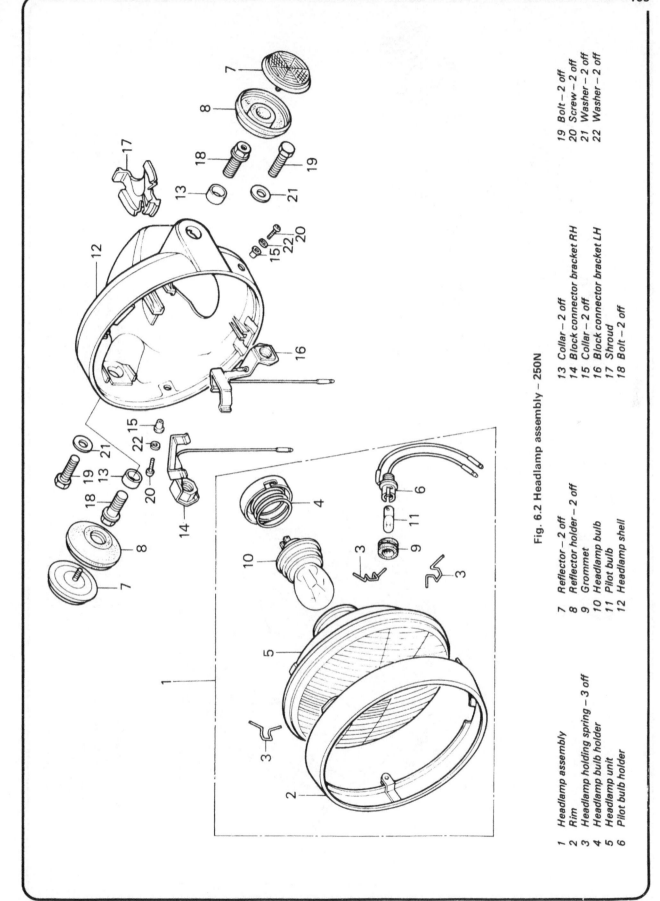

Fig. 6.2 Headlamp assembly – 250N

1 Headlamp assembly
2 Rim
3 Headlamp holding spring – 3 off
4 Headlamp bulb holder
5 Headlamp unit
6 Pilot bulb holder

7 Reflector – 2 off
8 Reflector holder – 2 off
9 Grommet
10 Headlamp bulb
11 Pilot bulb
12 Headlamp shell

13 Collar – 2 off
14 Block connector bracket RH
15 Collar – 2 off
16 Block connector bracket LH
17 Shroud
18 Bolt – 2 off

19 Bolt – 2 off
20 Screw – 2 off
21 Washer – 2 off
22 Washer – 2 off

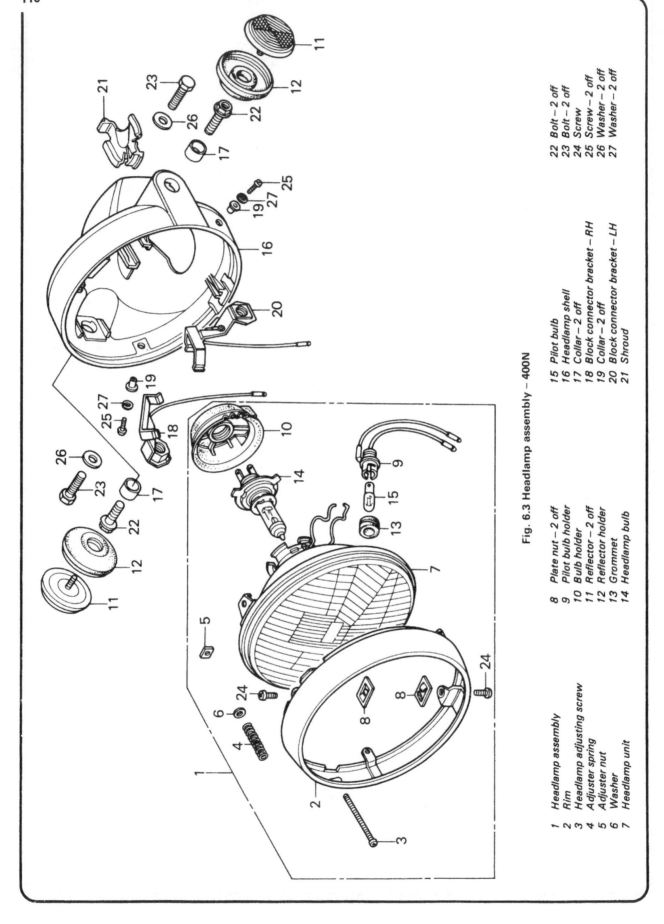

Fig. 6.3 Headlamp assembly – 400N

1 Headlamp assembly
2 Rim
3 Headlamp adjusting screw
4 Adjuster spring
5 Adjuster nut
6 Washer
7 Headlamp unit

8 Plate nut – 2 off
9 Pilot bulb holder
10 Bulb holder
11 Reflector – 2 off
12 Reflector holder
13 Grommet
14 Headlamp bulb

15 Pilot bulb
16 Headlamp shell
17 Collar – 2 off
18 Block connector bracket – RH
19 Collar – 2 off
20 Block connector bracket – LH
21 Shroud

22 Bolt – 2 off
23 Bolt – 2 off
24 Screw
25 Screw – 2 off
26 Washer – 2 off
27 Washer – 2 off

10 Stop and tail lamp: replacement of bulbs

1   The combined stop and tail lamp bulb contains two filaments, one for the stop lamp and one for tail lamp.
2   The offset pin bayonet fixing bulb can be renewed after the plastic lens cover and screws has been removed.

11 Flashing indicator lamps: replacing bulbs

1   Flashing indicator lamps are fitted to the front and rear of the machine. They are mounted on short stalks through which the wires pass. Access to each bulb is gained by removing the two screws holding the plastic lens cover. The bulbs are retained by a bayonet fixing.

11.1 ... each flashing indicator lens

10.1 Stop/tail lamp lens is held by two screws as is ...

12 Flashing indicator relay: location and replacement

1   The flashing indicator relay fitted in conjunction with the flashing indicator lamps is located behind the rectifier, behind the left-hand side cover. It is mounted in a rubber 'box' which isolates it from the harmful effects of vibration.
2   When the relay malfunctions, it must be renewed; a repair is impracticable. When the unit is in working order audible clicks will be heard which coincide with the flash of the indicator lamps. If the lamps malfunction, check firstly that a bulb has not blown, or the handlebar switch is not faulty. The usual symptom of a fault is one initial flash before the unit goes dead.
3   Take great care when handling a flasher unit. It is easily damaged, if dropped.

13 Warning and indicator lamps: bulb replacement

1   All bulbs fitted to the instrument heads or warning lamp console are of the bayonet fit type.
2   To gain access to the bulbs commence by removing the instrument console cover which is held at the periphery by screws. The central cluster of bulbs may be removed from their holders, one at a time. To allow removal of the instrument mounted bulbs, the drive cable on the instrument head must be detached and the two mounting nuts removed from the base of the casing. The instrument can then be lifted upward, sufficiently to allow the bulb holder in question to be pulled out.

14 Horn: location

1   The horn is mounted below the headlamp shell on a bracket secured by one fork yoke pinch bolt. In the event of malfunction, no repair or adjustment is possible and the horn must be renewed.

15 Ignition switch: removal and replacement

1   The combined ignition and lighting master switch is mounted in between the two instrument heads, secured to the instrument console cover. If the switch malfunctions it may be removed after detaching the cover and after disconnecting the leads at the block connector. Repair is rarely practicable. It is preferable to purchase a new switch unit, which will probably necessitate the use of a different key.

16 Stop lamp switch: adjustment

1   All models have a stop lamp switch fitted to operate in conjunction with the rear brake pedal. The switch is located immediately to the rear of the crankcase, on the right-hand side of the machine. It has a threaded body giving a range adjustment.
2   If the stop lamp is late in operating, turn the adjuster nut in a clockwise direction so that the switch rises from the bracket to which it is attached.
3   If the lamp operates too early, the adjuster nut should be turned anti-clockwise so that the switch body is lowered in relation to the mounting bracket.
4   As a guide, the light should operate after the brake pedal has been depressed by about 2 cm ($\frac{3}{4}$ inch).
5   A stop lamp switch is also incorporated in the front brake cable, to give warning when the front brake is applied. This is not yet a statutory requirement in the UK, although it applies in many other countries and states.
6   The front brake stop lamp switch is built into the hydraulic system and contains no provision for adjustment. If the switch malfunctions, it must be renewed.

17 Neutral indicator switch: location and testing

1 Both models have a neutral indicator switch mounted in the roof of the gearbox. The switch is interconnected with a warning bulb in the console between the instrument heads mounted forward of the handlebars.

2 If the neutral indicator lamp does not illuminate at the correct time, remove the boot from the top of the switch and disconnect the lead. With the ignition switched on, earth the lead against the engine casing. If the light does not come on, a faulty lead or bulb is indicated. If the light works, the switch is at fault.

18 Handlebar switches: general

1 Generally speaking, the switches give little trouble, but if necessary they can be dismantled by separating the halves which form a split clamp around the handlebars. Note that the machine cannot be started until the ignition cut-out on the right-hand end of the handlebars is turned to the central 'ON' position.

2 Always disconnect the battery before removing any of the switches, to prevent the possibility of short circuit. Most troubles are caused by dirty contacts, but in the event of the breakage of some internal part, it will be necessary to renew the complete switch.

3 Because the internal components of each switch are very small, and therefore difficult to dismantle and reassemble, it is suggested a special electrical contact cleaner be used to clean corroded contacts. This can be sprayed into each switch, without the need for dismantling.

19 Fault diagnosis: electrical system

Symptom	Cause	Remedy
Complete electrical failure	Blown fuse	Check wiring and electrical components for short circuit before fitting a new fuse. Check battery connections, also whether connections show signs of corrosion.
Dim lights, horn inoperative	Discharged battery	Recharge battery with battery charger and check whether alternator is giving correct output (electrical specialist).
Constantly 'blowing' bulbs	Vibration, poor earth connection	Check whether bulb holders are secured correctly. Check earth return or connections to frame.

1982 Honda CB400 NC

Chapter 7 Honda CB250 N and CB400 N Super Dreams 1981 to 1984 models

Contents

Specifications

Specifications below are given only where they differ from those for the earlier Super Dream models featured at the beginning of each of the first six Chapters of this manual.

Specifications relating to Chapter 2
Carburettor

	CB250 NDC	CB400 NC
Make	Keihin	Keihin
Type	VB30B	VB31B
Primary main jet	72	78
Secondary main jet	105	108
Slow jet	38	42
Air screw opening	2 turns	1½ turns
Float height	15.5 mm (0.61 in)	15.5 mm (0.61 in)
Idle speed	1300 ± 100 rpm	1200 ± 100 rpm

Specifications relating to Chapter 4
Front forks

CB250 NB, NDB, NDC and CB400 NB, NC

Type	Telescopic, hydraulically damped with bushed legs
Travel	140 mm (5.5 in)
Spring free length	499.2 mm (19.65 in)
Service limit	489.2 mm (19.26 in)
Stanchion bend (max)	0.2 mm (0.008 in)
Oil capacity (per leg)	190 cc (6.6 Imp fl oz)

Specifications relating to Chapter 5
Tyres

CB250 NB, NDB, NDC and CB400 NB, NC

Front	3.60S19-4PR tubeless
Rear	4.10S18-4PR tubeless

Tyre pressures

	Solo	Pillion
Front	28 psi	28 psi
Rear	28 psi	36 psi

Master cylinder

Master cylinder ID	14.00 – 14.043 mm (0.5512 – 0.5529 in)
Service limit	14.055 mm (0.5533 in)
Master cylinder piston OD:	
CB250 N	13.850 – 13.907 mm (0.5453 – 0.5475 in)
Service limit	13.835 mm (0.5447 in)
CB400 N	13.957 – 13.984 mm (0.5495 – 0.5506 in)
Service limit	13.945 mm (0.5490 in)

Brake caliper

CB400 NC

Caliper piston OD	30.150 – 30.200 mm (1.1870 – 1.1890 in)
Service limit	30.142 mm (1.1867 in)
Caliper bore ID	30.230 – 30.306 mm (1.1902 – 1.1931 in)
Service limit	30.136 mm (1.1865 in)

Specifications relating to Chapter 6
Headlamp bulb

CB250 NDC
60/55W

Alternator output	Charging starts	Output at 5000 rpm
CB250 NDC	1300 rpm min	15/5A min/14.5V
CB400 NC	1300 rpm min	12–14A min/14.5V

1 Introduction

Since 1978 when the Honda Super Dreams were introduced they have proved to be extremely popular, especially the 250 model which topped the motorcycle sales chart for a considerable time. There was obviously no need for Honda to change their models to attract greater sales. For this reason the later models, covered in this Chapter, have not been changed a great deal from their predecessors.

The most significant changes have been made to the cycle parts with the introduction of bushed fork legs, tubeless tyres and a twin piston caliper front brake, which is only fitted to the 1982 to 1984 CB400 N model. Minor engine modifications have been made; these are discussed in the text. As previously mentioned various small modifications have been made that do not necessarily affect the dismantling procedure; these relate to the front brake master cylinder reservoir and the brake disc mounting arrangement.

2 Camshaft lubrication pipe: removal and refitting – CB250 NDC and CB400 NC

1 The 1982 on models have been fitted with oil transfer pipes to carry extra lubrication to the camshaft. Oil is delivered to the camshaft by way of a single pipe from the crankcase to the cylinder head which branches into two pipes to feed each of the camshaft holders.
2 Little should go wrong with the system apart from possibly a blockage of the pipes or, more probably, oil leakage from the unions. Blockages can be cleared by removing each pipe and directing a blast of air through them.
3 The external pipe running down the centre of the cylinder head can be dismantled by removing the banjo union bolt from the top of the crankcase and the banjo union bolt from the top of the cylinder head. A sealing washer is fitted on each side of the banjo union and must always be refitted with the pipe.
4 Access can be gained to the branched oil pipe once the cylinder head cover is removed. Remove the two bolts and lift the cover off the cylinder head. The oil pipe can now be clearly seen. The oil delivery connection at the front of the head shares the same retaining bolt as the external pipe. An O-ring is fitted between the delivery pipe union and the cylinder head. Finally remove the two bolts holding the branches of the pipe to the camshaft holders.
5 When refitting either pipe ensure that the oilways in the banjo union are free of blockages and that the sealing washers are in good condition. Because the banjo bolts are drilled axially

and radially they have a low shear strength and are easily broken. Care in tightening should be adopted particularly as no torque figures are quoted by the manufacturer. Leakage should be overcome by renewing the union washers rather than further tightening of the banjo bolts.

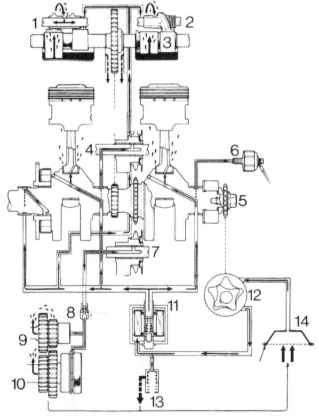

Fig. 7.1 Engine/gearbox lubrication system – CB250 NDC and CB400 NC models

1	Rocker arm	8	Orifice
2	Rocker arm	9	Mainshaft
3	Camshaft	10	Layshaft
4	Front balancer	11	Oil filter
5	Crankshaft	12	Oil pump
6	Oil pressure switch	13	Pressure relief valve
7	Rear balancer	14	Oil strainer

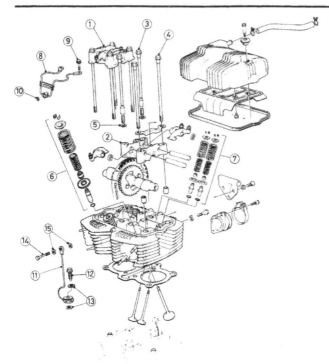

Fig. 7.2 Cylinder head and valve gear – CB250 NDC and CB400 NC models

1 Right-hand rocker carrier	7 Inlet valve assembly
2 Left-hand rocker carrier	8 Distributor pipe
3 Bolt with rubber seal – 2 off	9 Union bolt
4 Bolt – 6 off	10 O-ring
5 Sealing washer – 8 off	11 Oil feed pipe
6 Exhaust valve assembly	12 Union bolt
	13 Sealing washers
	14 Union bolt
	15 Sealing washers

3 Cylinder head removal: CB250 NDC and CB400 NC models

1   The procedure for removing the cylinder head remains virtually unchanged, with the exception of the two innermost retaining bolts of the left-hand rocker carrier. A rubber seal is fitted to the stem of each bolt and forms an important function in the cylinder head lubrication system. It is for this reason that each bolt must be refitted in its correct location upon reassembly and must never be transposed with the plain type bolts.

4 Removing the gearchange external components – CB250 NDC and CB400 NC

1   A modified selector drum change pin assembly has been fitted to the above models. The procedure for removing the gearchange shaft and stopper arm is, however, unchanged.
2   The change pins can be removed by removing the single bolt in the centre of the selector drum cam and lifting the cam away. Access can now be made to the five pins.
3   Upon reassembly remember to locate the raised pin on the end of the selector drum with the drilling on the rear face of the cam.

5 Front forks: general description

The most notable change to the front fork legs is the adoption of bushed legs. Two teflon coated bushes are fitted in each leg which prevent actual contact of the stanchion and lower leg. These bushes are replacement items and can be renewed if wear develops in the fork leg.

The fork top bolt design varies between the B and C models. The type fitted to the C model maintains tension directly on the fork spring whereas the B model top bolt acts as a plug and a second spring retaining ring is fitted into the top of the stanchion.

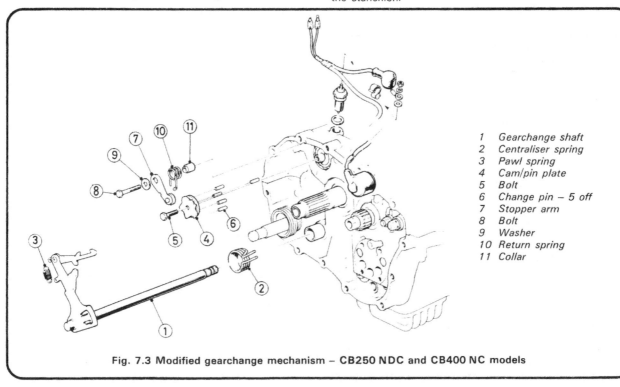

1	Gearchange shaft
2	Centraliser spring
3	Pawl spring
4	Cam/pin plate
5	Bolt
6	Change pin – 5 off
7	Stopper arm
8	Bolt
9	Washer
10	Return spring
11	Collar

Fig. 7.3 Modified gearchange mechanism – CB250 NDC and CB400 NC models

6 Front forks: removal, dismantling, examination and reassembly

1 The fork legs can be removed from the machine after the front wheel, mudguard and brake caliper(s) have been removed. Refer to Chapter 4, Section 2 for removal of these components.
2 Using an open ended spanner slacken and remove the fork top bolt from each leg. The top bolt fitted to the C model is under pressure from the spring and once the top bolt is removed the spring will extend forcibly. For this reason unscrew the bolt slowly applying an opposing force with the spanner so that the threads are not damaged as the plug leaves the last threads. Slacken the fork yoke upper and lower pinch bolts and pull the legs downwards out of their location. If any resistance is felt twist the leg as it is pulled from the machine, ensuring that the machine does not overbalance in the process.
3 If the forks are to be dismantled tackle each leg separately to avoid the possibility of interchanging the components.

B models only
4 Select an Allen key of the size of the recess in the spring retaining ring and proceed to unscrew the ring. This ring will be under spring tension and care must be taken to prevent the spring forcing the ring into orbit. It may be necessary to hold the stanchion in a vice to prevent its rotating whilst the retaining ring is slackened. Make sure that the jaws of the vice are padded and that only a small amount of force is applied to the stanchion. Remove the spring seat from the top of the spring.

C model
5 Remove the spring collar from the top of the stanchion.

All models
6 Invert the fork leg and allow the oil to drain. The fork spring can now be shaken out of the stanchion. As the spring emerges from the stanchion note the direction of the closer wound coils as a guide for reassembly. Prise off the rubber dust seal to reveal the circlip located beneath it. Using a pair of circlip pliers dislodge the circlip from its groove and slide it over the end of the stanchion.
7 In order to separate the stanchion from the lower leg the Allen bolt which resides in the base of the lower leg must first be removed. This bolt screws into the damper rod which is situated inside the stanchion and will often rotate with the damper rod when an attempt is made to unscrew it. Because of this provision must be made to hold the damper rod. The head of the damper rod is recessed to accommodate a service tool which can be obtained through a Honda dealer. This tool is a length of steel bar with a head that will locate in the top of the rod. Alternatively the following methods can be utilised to the same effect.
8 Place the lower leg of the fork between the padded jaws of a vice and tighten the vice just enough to hold the leg in place. Temporarily install the fork spring and fit the top cap. This method uses the force of the compressed fork spring to prevent the damper rod from rotating. An attempt can then be made to unscrew the Allen bolt. If this method proves unsatisfactory obtain a length of wooden dowelling, slightly taper one end and install the tapered end in the head of the damper rod. With the aid of an assistant the dowelling protruding from the stanchion can be held firm whilst the Allen bolt is unscrewed. If an assistant is not available cut the dowelling off to just below the top of the stanchion and refit the fork top bolt.
9 With the Allen bolt removed the two components can be separated by using the stanchion as a slide hammer by which to dislodge the bush, together with the oil seal and backing rings. Once the two components are separated the damper rod seat can be shaken out of the lower leg.
10 Clean all the components and lay them out on a clean work

surface and examine them for wear and damage as follows.
11 It is generally not possible to straighten forks which have been damaged in an accident, particularly when the correct jigs are not available. It is always best to err on the side of safety and fit new ones, especially since there is no easy means to detect whether the forks have been overstressed or metal fatigued. Fork stanchions can be checked, after removal from the lower legs, by rolling them on a dead flat surface. Any misalignment will be immediately obvious.
12 Check the free length of the fork springs and replace if they exceed the minimum length of 489 mm (19.25 in). Fork springs will take a permanent set after considerable usage and will require renewal if the fork action becomes spongy.
13 Visually inspect the bushes for signs of scoring and chipping. It is good practice to renew the bushes if the teflon surface coating has worn to such an extent that the copper bush is showing for more than three-quarters of the entire surface.
14 Check the condition of the rubber dust seals. If they are cracked or worn around the stanchion area dust and water will penetrate the seal and eventually damage the oil seal underneath. It is advisable to renew the oil seals when the forks are dismantled, even if they appear to be in good condition. This will save a strip-down of the fork at a later date if oil leakage occurs.
15 If damping action is lost, the piston ring around the damper rod head should be renewed. Clear any obstructions from the small holes in the rod and check the action of the short rebound spring.
16 When reassembling the fork leg conduct reassembly in clean working conditions and ensure that all components are absolutely clean. Position the short rebound spring over the end of the damper rod and insert the assembly into the stanchion. Fit the damper rod seat over the end of the protruding damper rod and carefully insert the stanchion into the lower leg. Coat the threads of the Allen bolt with a locking compound and insert the bolt and its washer into the end of the damper rod. One of the methods described for slackening the assembly can be used for tightening it.
17 The upper bush, oil seal and backing rings should now be fitted. Position the bush over the stanchion and slide it along until it rests against its housing. It is necessary to use either a fabricated tool or the Honda service tool to drive the bush into the lower leg recess. It is important that the bush is driven in squarely so that damage to the fork leg is not incurred. To fabricate the home-made tool obtain a length of tube of an internal diameter slightly larger than that of the stanchion and an external diameter slightly less than that of the bush. Ensure that the end of this tube is both square to its sides and free of burrs. By sliding this tube up and down the stanchion an even force can be placed on the outboard surface of the bush. Slide the lower backing ring over the stanchion and install it against the bush. Lubricate the oil seal in ATF and fit it on top of the backing ring, with its marked face uppermost. The tool used for refitting of the bush may also be used to drive the oil seal home. Note that the oil seal is fitted correctly when the circlip groove around the inside of the housing is clearly visible. Install the upper backing ring over the oil seal and secure it with the circlip. Finally refit the dust seal ensuring that it locates properly over the lower leg.
18 Refill the fork leg and install the spring with its closer coils positioned in the same direction as noted during reassembly.

B models
19 Replace the spring seat and the spring retaining ring. Clamp the stanchion between the padded jaws of a vice while the ring is tightened. Then fit the cap over the ring.

C model
20 Position the spring collar on top of the spring.

All models

21 Refit the fork legs into their yokes so that the top edge of the stanchion is flush with the face of the upper yoke and tighten the yoke pinch bolts.

22 Inspect the O-ring around the fork top bolt and renew it if necessary. Refit the top bolt into the stanchion. Refit the front mudguard, wheel and brake caliper(s) and test the action of the forks by bouncing them up and down.

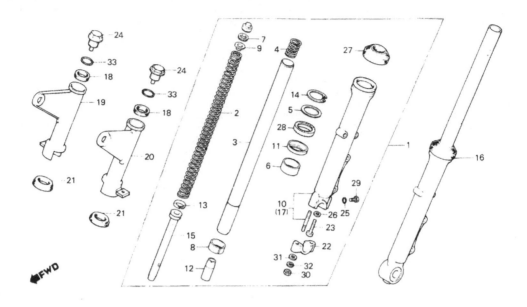

Fig. 7.4 Front forks – B models (C models similar)

1	Right-hand fork leg assembly	9 Spring seat	17 Stud – 2 off	24 Top bolt
2	Spring	10 Lower leg	18 Damping ring – 2 off	25 Sealing washer
3	Stanchion	11 Lower backing ring	19 Right-hand headlamp bracket	26 Sealing washer
4	Rebound spring	12 Damper rod seat	20 Left-hand headlamp bracket	27 Dust seal
5	Upper backing ring	13 Damper rod piston ring	21 Damping ring – 2 off	28 Oil seal
6	Upper bush	14 Circlip	22 Wheel spindle clamp	29 Drain bolt
7	Spring retaining ring	15 Damper rod	23 Allen bolt	30 Nut – 2 off
8	Lower bush	16 Left-hand fork leg		31 Washer – 2 off
			32 Spring washer – 2 off

---

7 Front brake caliper: dismantling, examination and re-assembly – CB400 NC

1 Brake pad wear can be checked by looking through the aperture at the rear of the caliper. The caliper can remain in situ for this operation. Replace the pads if either has worn down to the wear line scribed on the top edge. Replace brake pads as a set, never singly.

2 The brake pads are located inside the caliper by two pins which must be removed prior to pad removal. Remove the single bolt at the rear of the caliper, which retains the pin retaining plate and lift off the plate. Unscrew the lower of the two bolts which secure the caliper body to the caliper support bracket and pivot the caliper up, away from the bracket and disc. Using a pair of long-nose pliers pull the pad retaining pins from position. The brake pads can now be removed for inspection or replacement.

3 When refitting the brake pads ensure that the spring plate is positioned correctly in the caliper base. Due to the increased thickness of the new pads the pistons may have to be pushed back into their bore. This will require a firm pressure and must only be done with the hand, do not strike the piston at all. Refit the pads and install the locating pins. Reassembly of the caliper is a direct reversal of the dismantling sequence.

4 If attention to the caliper internal components is required the system must first be drained of fluid and the caliper removed from the machine. Place a suitable receptacle beneath the brake hose banjo bolt on the caliper, and remove the banjo

bolt. Take great care not to allow hydraulic fluid to spill onto paintwork; it is a very effective paint stripper. Hydraulic fluid will also damage rubber and plastic components.

5 Remove the caliper and brake pads as described earlier and lift out the brake pad spring plate. Prise out the collar and rubber boot from the lug situated above the piston housings. The pistons can be withdrawn from their bores by applying a small amount of compressed air through the fluid inlet. Place a piece of rag over the projecting piston heads to absorb the fluid still in the caliper. If the piston oil seals require renewal they can be prised from position with the flat blade of a screwdriver. Care must be taken not to damage the bore when removing the seal.

6 Wash all internal components in new hydraulic fluid and examine them for wear or deterioration as follows.

7 Check the piston surfaces and the caliper bore for scoring or excessive wear. It is a good idea to renew all seals whilst the caliper is apart, this will obviate the risk of a seal failing shortly after the caliper is in operation again. Needless to say, if a particular seal is damaged it must certainly be renewed.

8 Coat the pistons and inner lips of the seals with new brake fluid and install them in the caliper. Assemble the caliper components by reversing the dismantling procedure.

9 Refit the caliper to the machine and reconnect the hydraulic hose and banjo union bolt. Wipe all traces of fluid from the caliper exterior and commence to bleed the system of air. Place a clean container under the caliper and connect a length of plastic pipe to the bleed nipple and the container. Remove the two screws retaining the master cylinder reservoir cap and add hydraulic fluid up to the upper level line. Loosely refit the cap.

Close the bleed nipple and pump the brake lever until the new fluid has travelled through the brake line. Keep a close watch on the level of fluid in the reservoir making sure that it does not fall below the bottom level.

10 Unscrew the bleed screw one half turn and squeeze the brake lever as far as it will go but do not release it until the

bleeder valve is closed again. Repeat the operation a few times until no more air bubbles come from the plastic tube.

11 Keep topping up the reservoir with new fluid. When all the bubbles disappear, close the bleeder valve dust cap. Check the fluid level in the reservoir after the bleeding operation has been completed.

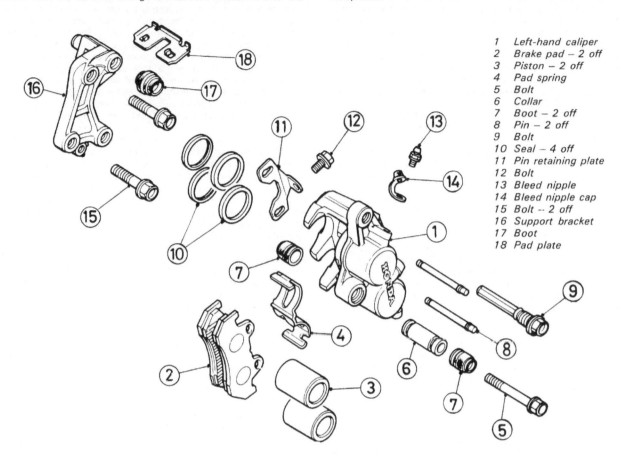

1   Left-hand caliper
2   Brake pad – 2 off
3   Piston – 2 off
4   Pad spring
5   Bolt
6   Collar
7   Boot – 2 off
8   Pin – 2 off
9   Bolt
10 Seal – 4 off
11 Pin retaining plate
12 Bolt
13 Bleed nipple
14 Bleed nipple cap
15 Bolt -- 2 off
16 Support bracket
17 Boot
18 Pad plate

Fig. 7.5 Front brake caliper – CB400 NC model

8 Front brake disc: removal and replacement – CB250 NDB, NDC and CB400 NC

CB250 NDB and NDC

1 In order to remove the brake disc the front wheel and speedometer cable should first be removed as described in Chapter 5, Section 3.

2 Lay the wheel flat on the ground, with the hub cover uppermost and undo the five nuts to free the cover. The five long bolts can then be pulled through the left-hand side of the wheel to free the brake disc.

3 Refitting of the disc is a straightforward reversal of the removal procedure. Tighten the nuts to a torque setting of 2.7 – 3.3 kgf m (20 – 24 lbf ft).

CB400 NC

4 Remove the front wheel and speedometer cable as described in Chapter 5, Section 3.

5 Both discs are retained by five long bolts which pass directly through the hub from the left-hand side, and are held by five nuts on the opposite side of the wheel. Note that a damping spacer is fitted immediately underneath each disc.

6 Reassembly is a straightforward reversal of the dismantling procedure. Tighten the nuts to a torque setting of 2.7 – 3.3 kgf m (20 – 24 lbf ft).

9 Tyres: removal and refitting (tubeless tyres)

1 It is strongly recommended that should a repair to a tubeless tyre be necessary, the wheel is removed from the machine and taken to a tyre fitting specialist who is willing to do the job or taken to an official dealer. This is because the force required to break the seal between the wheel rim and tyre bead is considerable and considered to be beyond the capabilities of an individual working with normal tyre removing tools. Any abortive attempt to break the rim to bead seal may also cause damage to the wheel rim, resulting in an expensive wheel replacement. If, however, a suitable bead releasing tool is available, and experience has already been gained in its use, tyre removal and refitting can be accomplished as follows.

2 Remove the wheel from the machine by following the instructions for wheel removal as described in Chapter 5, Section 3 or 11, depending upon which wheel is involved. Deflate the tyre by removing the valve insert and when it is fully deflated, push the bead of the tyre away from the wheel rim on both sides so that the bead enters the centre well of the rim. As noted, this operation will almost certainly require the use of a bead releasing tool.

3 Insert a tyre lever close to the valve and lever the edge of the tyre over the outside of the wheel rim. Very little force should be necessary; if resistance is encountered it is probably

due to the fact that the tyre beads have not entered the well of the wheel rim all the way round the tyre. Should the initial problem persist, lubrication of the tyre bead and the inside edge and lip of the rim will facilitate removal. Use a recommended lubricant, a diluted solution of washing-up liquid or french chalk. Lubrication is usually recommended as an aid to tyre fitting but its use is equally desirable during removal. The risk of lever damage to wheel rims can be minimised by the use of proprietary plastic rim protectors placed over the rim flange at the point where the tyre levers are inserted. Suitable rim projectors may be fabricated very easily from short lengths (4-6 inches) of thick-walled nylon petrol pipe which have been split down one side using a sharp knife. The use of rim protectors should be adopted whenever levers are used and, therefore, when the risk of damage is likely.

4   Once the tyre has been edged over the wheel rim, it is easy to work around the wheel rim so that the tyre is completely free on one side.

5   Working from the other side of the wheel, ease the other edge of the tyre over the outside of the wheel rim, which is furthest away. Continue to work around the rim until the tyre is freed completely from the rim.

6   Refer to the following Section for details relating to puncture repair and the renewal of tyres. See also the remarks relating to the tyre valves in Section 11.

7   Refitting of the tyre is virtually a reversal of removal procedure. If the tyre has a balance mark (usually a spot of coloured paint), as on the tyres fitted as original equipment, this must be positioned alongside the valve. Similarly, any arrow indicating direction of rotation must face the right way.

8   Starting at the point furthest from the valve, push the tyre bead over the edge of the wheel rim until it is located in the central well. Continue to work around the tyre in this fashion until the whole of one side of the tyre is on the rim. It may be necessary to use a tyre lever during the final stages. Here again, the use of a lubricant will aid fitting. It is recommended strongly that when refitting the tyre only a recommended lubricant is used because such lubricants also have sealing properties. Do not be over generous in the application of lubricant or tyre creep may occur.

9   Fitting the upper bead is similar to fitting the lower bead. Start by pushing the bead over the rim and into the well at a point diametrically opposite the tyre valve. Continue working round the tyre, each side of the starting point, ensuring that the bead opposite the working area is always in the well. Apply lubricant as necessary. Avoid using tyre levers unless absolutely essential, to help reduce damage to the soft wheel rim. The use of the levers should be required only when the final portion of bead is to be pushed over the rim.

10   Lubricate the tyre beads again prior to inflating the tyre, and check that the wheel rim is evenly positioned in relation to the tyre beads. Inflation of the tyre may well prove impossible without the use of a high pressure air hose. The tyre will retain air completely only when the beads are firmly against the rim edges at all points and it may be found when using a foot pump that air escapes at the same rate as it is pumped in. This problem may also be encountered when using an air hose on new tyres which have been compressed in storage and by virtue of their profile hold the beads away from the rim edges. To overcome this difficulty, a tourniquet may be placed around the circumference of the tyre, over the central area of the tread. The compression of the tread in this area will cause the beads to be pushed outwards in the desired direction. The type of tourniquet most widely used consists of a length of hose closed at both ends with a suitable clamp fitted to enable both ends to be connected. An ordinary tyre valve is fitted at one end of the tube so that after the hose has been secured around the tyre it may be inflated, giving a constricting effect. Another possible method of seating beads to obtain initial inflation is to press the tyre into the angle between a wall and the floor. With the airline attached to the valve additional pressure is then applied to the tyre by the hand and shin, as shown in the accompanying illustration. The application of pressure at four points around the tyre's circumference whilst simultaneously applying the air hose will often effect an initial seal between the tyre beads and wheel rim, thus allowing inflation to occur.

11   Having successfully accomplished inflation, increase the pressure to 40 psi and check that the tyre is evenly disposed on the wheel rim. This may be judged by checking that the thin positioning line found on each tyre wall is equidistant from the rim around the total circumference of the tyre. If this is not the case, deflate the tyre, apply additional lubrication and reinflate. Minor adjustments to the tyre position may be made by bouncing the wheel on the ground.

12   Always run the tyre at the recommended pressures and never under or over-inflate. The correct pressures are given in the Specifications Section of this Chapter.

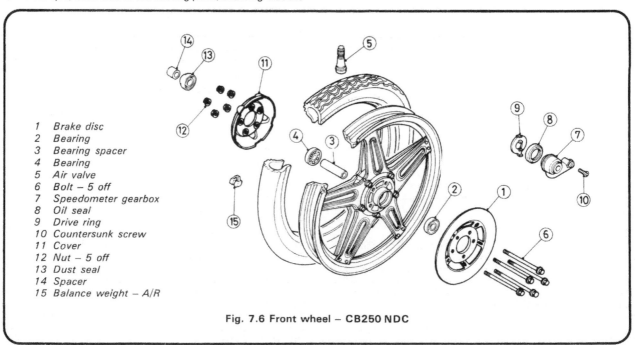

1   Brake disc
2   Bearing
3   Bearing spacer
4   Bearing
5   Air valve
6   Bolt – 5 off
7   Speedometer gearbox
8   Oil seal
9   Drive ring
10 Countersunk screw
11 Cover
12 Nut – 5 off
13 Dust seal
14 Spacer
15 Balance weight – A/R

Fig. 7.6 Front wheel – CB250 NDC

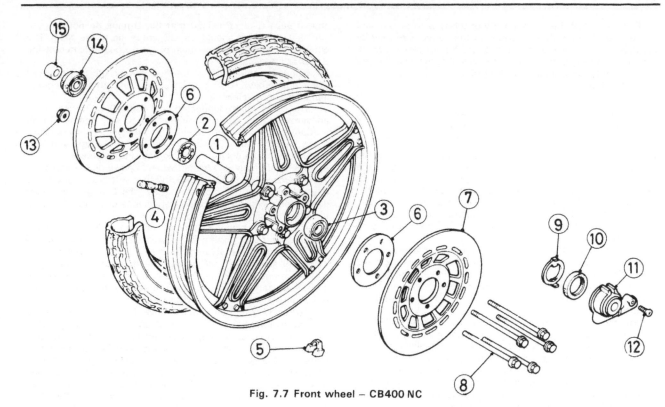

Fig. 7.7 Front wheel – CB400 NC

1 Bearing spacer	4 Air valve	7 Brake disc	10 Oil seal	13 Nut – 5 off
2 Bearing	5 Balance weight – A/R	8 Bolt – 5 off	11 Speedometer gearbox	14 Dust seal
3 Bearing	6 Damping spacer – 2 off	9 Drive dog	12 Countersunk screw	15 Spacer

HAND

WALL

SHIN

FLOOR

H11400

Fig. 7.8 Method of seating beads of tubeless tyres

10 Puncture repair and tyre renewal

1   The primary advantage of the tubeless tyre is its ability to accept penetration by sharp objects such as nails etc without loss of air. Even if loss of air is experienced, because there is no inner tube to rupture, in normal conditions a sudden blow-out is avoided.

2   If a puncture of the tyre occurs, the tyre should be removed for inspection for damage before any attempt is made at remedial action. The temporary repair of a punctured tyre by inserting a plug from the outside should not be attempted. Although this type of temporary repair is used widely on cars, the manufacturers strongly recommend that no such repair is carried out on a motorcycle tyre. Not only does the tyre have a thinner carcass, which does not give sufficient support to the plug, the consequences of a sudden deflation are often sufficiently serious that the risk of such an occurrence should be avoided at all costs.

3   The tyre should be inspected both inside and out for damage to the carcass. Unfortunately the inner lining of the tyre – which takes the place of the inner tube – may easily obscure any damage and some experience is required in making a correct assessment of the tyre condition.

4   There are two main types of tyre repair which are considered safe for adoption in repairing tubeless motocycle tyres. The first type of repair consists of inserting a mushroom-headed plug into the hole from the inside of the tyre. The hole is prepared for insertion of the plug by reaming and the application of an adhesive. The second repair is carried out by buffing the inner lining in the damaged area and applying a cold or vulcanised patch. Because both inspection and repair, if they are to be carried out safely, require experience in this type of work, it is recommended that the tyre be placed in the hands of a repairer with the necessary skills, rather than repaired in the home workshop.

5   In the event of an emergency, the only recommended 'get-you-home' repair is to fit a standard inner tube of the correct size. If this course of action is adopted, care should be taken to ensure that the cause of the puncture has been removed before the inner tube is fitted. it will be found that the valve hole in the rim is considerably larger than the diameter of the inner tube valve stem. To prevent the ingress of road dirt, and to help support the valve, a spacer should be fitted over the valve. A conversion spacer for most Honda models equipped with Comstar wheels is available from Honda dealers.

6   In the event of the unavailability of tubeless tyres, ordinary tubed tyres fitted with inner tubes of the correct size may be fitted. Refer to the manufacturer or a tyre fitting specialist to ensure that only a tyre and tube of equivalent type and suitability is fitted, and also to advise on the fitting of a valve nut/spacer to the rim hole.

11 Tyre valves: description and renewal

1   It will be appreciated from the preceding Sections, that the adoption of tubeless tyres has made it necessary to modify the valve arrangement, as there is no longer an inner tube which can carry the valve core. The problem has been overcome by using a moulded rubber valve body which locates in the wheel rim hole. The valve body is pear-shaped, and has a groove around its widest point which engages with the rim forming an airtight seal.

2   The valve is fitted from the rim well, and it follows that it can only be renewed whilst the tyre itself is removed from the wheel. Once the valve has been fitted, it is almost impossible to remove it without damage, and so the simplest method is to cut it as close as possible to the rim well. The two halves of the old valve can then be removed.

3   The new valve is fitted by inserting the threaded end of the valve body through the rim hole, and pulling it through until the groove engages in the rim. In practice, a considerable amount of pressure is required to pull the valve into position, and most tyre fitters have a special tool which screws onto the valve end to enable purchase to be obtained. It is advantageous to apply a little tyre bead lubricant to the valve to ease its insertion. Check that the valve is seated evenly and securely.

4   The incidence of valve body failure is relatively small, and leakage only occurs, when the rubber valve case ages and begins to perish. As a precautionary measure, it is advisable to fit a new valve when a new tyre is fitted. This will preclude any risk of the valve failing in service. When purchasing a new valve, it should be noted that a number of different types are available. The correct type for use in the Comstar wheel is a Schrader 413, Bridgeport 183M or equivalent.

5   The valve core is of the same type as that used with tubed tyres and screws into the valve body. The core can be removed with a small slotted tool which is normally incorporated in plunger type pressure gauges. Some valve dust caps incorporate a projection for removing valve cores. Although tubeless tyre valves seldom give trouble, it is possible for a leak to develop if a small particle of grit lodges on the sealing face. Occasionally, an elusive slow puncture can be traced to a leaking valve core, and this should be checked before a genuine puncture is suspected.

6   The valve dust caps are a significant part of the tyre valve assembly. Not only do they prevent the ingress of road dirt into the valve, but also act as a secondary seal which will reduce the risk of sudden deflation if a valve core should fail.

12 O-ring final drive chain: maintenance

1   All late models are fitted with a final drive chain of the O-ring type, in which grease is packed into the bearing surfaces and sealed by small O-rings set between the sideplates at the end of each roller. Provided that the O-rings do not fail, the lubricant contained should last the entire life of the chain. The outer surfaces of the chain must of course be lubricated at the specified intervals.

2   Honda recommend that the final drive chain be inspected every 600 miles (1000 km) for correct adjustment, adequate lubricant and be cleaned.

3   The chain can be lubricated with SAE 80 or 90 gear oil or an aerosol chain lubricant suitable for O-ring chains. On no account use lubricants suitable for ordinary chains because these will inevitably cause damage to the O-rings, and also never subject the chain to a chain bath in hot grease.

4   The chain must be cleaned when in place on the machine, using kerosene (paraffin) applied with a soft brush; never use any strong solvents which may attack the O-rings, and never use a steam cleaner or pressure washer to clean the chain, as these will damage the O-rings.

5   A chain wear label is stuck to the outer faces of the chain adjusting brackets and must be replaced with another label if the chain is renewed. Its purpose is to indicate chain wear. The label consists of a red and green zone. If the green zone aligns with the end of the swinging arm the chain should be considered serviceable, but when the end of the arm is in the red zone the chain must be renewed.

6   To renew the chain either remove the rear wheel or move it forwards until the chain can be detached from the sprocket. Remove the gearchange pedal linkage and the left-hand exhaust silencer, then remove the left-hand footrest mounting plate, which is secured by two bolts and the swinging arm pivot shaft retaining nut. Remove the engine left-hand cover, then unscrew the gearbox sprocket mounting bolts, withdraw the retaining plate and pull the sprocket and chain off the shaft splines. Remove the left-hand rear suspension unit bottom mounting bolt, fully slacken the top mounting bolt and pull the unit off its swinging arm lug. Remove its two mounting bolts and withdraw the chainguard from the swinging arm. Withdraw the chain from the machine.

7   Refitting is the reverse of the removal procedure, ensuring that all fasteners are tightened to the specified torque wrench settings. Note that this procedure is only necessary for the standard endless chain which is supplied as a single loop or with a master link which must be rivetted to join both ends. However, O-ring chains are now available with ordinary connecting links: if one of these is fitted, the chain can be disconnected as described in Chapter 5 and run off the sprockets with no need for further dismantling. On refitting, ensure that the four O-rings are positioned correctly on each side of the rollers so that there is no space between the connecting link sideplate and spring clip. The closed end of the spring clip must face in the direction of chain travel.

8   Chain adjustment is carried out with the machine on its centre stand; the recommended free play is now 15 – 25 mm ($\frac{5}{8}$ – 1 in).

13 Headlamp: CB250 NDC

1   The Honda CB250 NDC is fitted with the same quartz halogen headlamp unit as the 400 model. Refer to Chapter 6, Section 9 and the figure 6.2 for headlamp removal. Remember that because of the nature of the quartz halogen bulb, it must not be handled.

Wiring diagram – Honda CB250 N and 400 N Super Dreams – All models

Conversion factors

Length (distance)

Inches (in)	X	25.4	= Millimetres (mm)	X 0.0394	= Inches (in)
Feet (ft)	X	0.305	= Metres (m)	X 3.281	= Feet (ft)
Miles	X	1.609	= Kilometres (km)	X 0.621	= Miles

Volume (capacity)

Cubic inches (cu in; in³)	X	16.387	= Cubic centimetres (cc; cm³)	X 0.061	= Cubic inches (cu in; in³)
Imperial pints (Imp pt)	X	0.568	= Litres (l)	X 1.76	= Imperial pints (Imp pt)
Imperial quarts (Imp qt)	X	1.137	= Litres (l)	X 0.88	= Imperial quarts (Imp qt)
Imperial quarts (Imp qt)	X	1.201	= US quarts (US qt)	X 0.833	= Imperial quarts (Imp qt)
US quarts (US qt)	X	0.946	= Litres (l)	X 1.057	= US quarts (US qt)
Imperial gallons (Imp gal)	X	4.546	= Litres (l)	X 0.22	= Imperial gallons (Imp gal)
Imperial gallons (Imp gal)	X	1.201	= US gallons (US gal)	X 0.833	= Imperial gallons (Imp gal)
US gallons (US gal)	X	3.785	= Litres (l)	X 0.264	= US gallons (US gal)

Mass (weight)

Ounces (oz)	X	28.35	= Grams (g)	X 0.035	= Ounces (oz)
Pounds (lb)	X	0.454	= Kilograms (kg)	X 2.205	= Pounds (lb)

Force

Ounces-force (ozf; oz)	X	0.278	= Newtons (N)	X 3.6	= Ounces-force (ozf; oz)
Pounds-force (lbf; lb)	X	4.448	= Newtons (N)	X 0.225	= Pounds-force (lbf; lb)
Newtons (N)	X	0.1	= Kilograms-force (kgf; kg)	X 9.81	= Newtons (N)

Pressure

Pounds-force per square inch (psi; lbf/in²; lb/in²)	X	0.070	= Kilograms-force per square centimetre (kgf/cm²; kg/cm²)	X 14.223	= Pounds-force per square inch (psi; lbf/in²; lb/in²)
Pounds-force per square inch (psi; lbf/in²; lb/in²)	X	0.068	= Atmospheres (atm)	X 14.696	= Pounds-force per square inch (psi; lbf/in²; lb/in²)
Pounds-force per square inch (psi; lbf/in²; lb/in²)	X	0.069	= Bars	X 14.5	= Pounds-force per square inch (psi; lbf/in²; lb/in²)
Pounds-force per square inch (psi; lbf/in²; lb/in²)	X	6.895	= Kilopascals (kPa)	X 0.145	= Pounds-force per square inch (psi; lbf/in²; lb/in²)
Kilopascals (kPa)	X	0.01	= Kilograms-force per square centimetre (kgf/cm²; kg/cm²)	X 98.1	= Kilopascals (kPa)
Millibar (mbar)	X	100	= Pascals (Pa)	X 0.01	= Millibar (mbar)
Millibar (mbar)	X	0.0145	= Pounds-force per square inch (psi; lbf/in²; lb/in²)	X 68.947	= Millibar (mbar)
Millibar (mbar)	X	0.75	= Millimetres of mercury (mmHg)	X 1.333	= Millibar (mbar)
Millibar (mbar)	X	0.401	= Inches of water (inH₂O)	X 2.491	= Millibar (mbar)
Millimetres of mercury (mmHg)	X	0.535	= Inches of water (inH₂O)	X 1.868	= Millimetres of mercury (mmHg)
Inches of water (inH₂O)	X	0.036	= Pounds-force per square inch (psi; lbf/in²; lb/in²)	X 27.68	= Inches of water (inH₂O)

Torque (moment of force)

Pounds-force inches (lbf in; lb in)	X	1.152	= Kilograms-force centimetre (kgf cm; kg cm)	X 0.868	= Pounds-force inches (lbf in; lb in)
Pounds-force inches (lbf in; lb in)	X	0.113	= Newton metres (Nm)	X 8.85	= Pounds-force inches (lbf in; lb in)
Pounds-force inches (lbf in; lb in)	X	0.083	= Pounds-force feet (lbf ft; lb ft)	X 12	= Pounds-force inches (lbf in; lb in)
Pounds-force feet (lbf ft; lb ft)	X	0.138	= Kilograms-force metres (kgf m; kg m)	X 7.233	= Pounds-force feet (lbf ft; lb ft)
Pounds-force feet (lbf ft; lb ft)	X	1.356	= Newton metres (Nm)	X 0.738	= Pounds-force feet (lbf ft; lb ft)
Newton metres (Nm)	X	0.102	= Kilograms-force metres (kgf m; kg m)	X 9.804	= Newton metres (Nm)

Power

Horsepower (hp)	X	745.7	= Watts (W)	X 0.0013	= Horsepower (hp)

Velocity (speed)

Miles per hour (miles/hr; mph)	X	1.609	= Kilometres per hour (km/hr; kph)	X 0.621	= Miles per hour (miles/hr; mph)

Fuel consumption*

Miles per gallon, Imperial (mpg)	X	0.354	= Kilometres per litre (km/l)	X 2.825	= Miles per gallon, Imperial (mpg)
Miles per gallon, US (mpg)	X	0.425	= Kilometres per litre (km/l)	X 2.352	= Miles per gallon, US (mpg)

Temperature

Degrees Fahrenheit = (°C x 1.8) + 32   Degrees Celsius (Degrees Centigrade; °C) = (°F - 32) x 0.56

*It is common practice to convert from miles per gallon (mpg) to litres/100 kilometres (l/100km), where mpg (Imperial) x l/100 km = 282 and mpg (US) x l/100 km = 235

Metric conversion tables

Inches	Decimals	Millimetres	Millimetres to Inches		Inches to Millimetres	
			mm	Inches	Inches	mm
1/64	0.015625	0.3969	0.01	0.00039	0.001	0.0254
1/32	0.03125	0.7937	0.02	0.00079	0.002	0.0508
3/64	0.046875	1.1906	0.03	0.00118	0.003	0.0762
1/16	0.0625	1.5875	0.04	0.00157	0.004	0.1016
5/64	0.078125	1.9844	0.05	0.00197	0.005	0.1270
3/32	0.09375	2.3812	0.06	0.00236	0.006	0.1524
7/64	0.109375	2.7781	0.07	0.00276	0.007	0.1778
1/8	0.125	3.1750	0.08	0.00315	0.008	0.2032
9/64	0.140625	3.5719	0.09	0.00354	0.009	0.2286
5/32	0.15625	3.9687	0.1	0.00394	0.01	0.254
11/64	0.171875	4.3656	0.2	0.00787	0.02	0.508
3/16	0.1875	4.7625	0.3	0.01181	0.03	0.762
13/64	0.203125	5.1594	0.4	0.01575	0.04	1.016
7/32	0.21875	5.5562	0.5	0.01969	0.05	1.270
15/64	0.234375	5.9531	0.6	0.02362	0.06	1.524
1/4	0.25	6.3500	0.7	0.02756	0.07	1.778
17/64	0.265625	6.7469	0.8	0.03150	0.08	2.032
9/32	0.28125	7.1437	0.9	0.03543	0.09	2.286
19/64	0.296875	7.5406	1	0.03937	0.1	2.54
5/16	0.3125	7.9375	2	0.07874	0.2	5.08
21/64	0.328125	8.3344	3	0.11811	0.3	7.62
11/32	0.34375	8.7312	4	0.15748	0.4	10.16
23/64	0.359375	9.1281	5	0.19685	0.5	12.70
3/8	0.375	9.5250	6	0.23622	0.6	15.24
25/64	0.390625	9.9219	7	0.27559	0.7	17.78
13/32	0.40625	10.3187	8	0.31496	0.8	20.32
27/64	0.421875	10.7156	9	0.35433	0.9	22.86
7/16	0.4375	11.1125	10	0.39370	1	25.4
29/64	0.453125	11.5094	11	0.43307	2	50.8
15/32	0.46875	11.9062	12	0.47244	3	76.2
31/64	0.484375	12.3031	13	0.51181	4	101.6
1/2	0.5	12.7000	14	0.55118	5	127.0
33/64	0.515625	13.0969	15	0.59055	6	152.4
17/32	0.53125	13.4937	16	0.62992	7	177.8
35/64	0.546875	13.8906	17	0.66929	8	203.2
9/16	0.5625	14.2875	18	0.70866	9	228.6
37/64	0.578125	14.6844	19	0.74803	10	254.0
19/32	0.59375	15.0812	20	0.78740	11	279.4
39/64	0.609375	15.4781	21	0.82677	12	304.8
5/8	0.625	15.8750	22	0.86614	13	330.2
41/64	0.640625	16.2719	23	0.09551	14	355.6
21/32	0.65625	16.6687	24	0.94488	15	381.0
43/64	0.671875	17.0656	25	0.98425	16	406.4
11/16	0.6875	17.4625	26	1.02362	17	431.8
45/64	0.703125	17.8594	27	1.06299	18	457.2
23/32	0.71875	18.2562	28	1.10236	19	482.6
47/64	0.734375	18.6531	29	1.14173	20	508.0
3/4	0.75	19.0500	30	1.18110	21	533.4
49/64	0.765625	19.4469	31	1.22047	22	558.8
25/32	0.78125	19.8437	32	1.25984	23	584.2
51/64	0.796875	20.2406	33	1.29921	24	609.6
13/16	0.8125	20.6375	34	1.33858	25	635.0
53/64	0.828125	21.0344	35	1.37795	26	660.4
27/32	0.84375	21.4312	36	1.41732	27	685.8
55/64	0.859375	21.8281	37	1.4567	28	711.2
7/8	0.875	22.2250	38	1.4961	29	736.6
57/64	0.890625	22.6219	39	1.5354	30	762.0
29/32	0.90625	23.0187	40	1.5748	31	787.4
59/64	0.921875	23.4156	41	1.6142	32	812.8
15/16	0.9375	23.8125	42	1.6535	33	838.2
61/64	0.953125	24.2094	43	1.6929	34	863.6
31/32	0.96875	24.6062	44	1.7323	35	889.0
63/64	0.984375	25.0031	45	1.7717	36	914.4

Index